BIORHYTHMS

A practical guide to using the new science of
biorhythms in everyday life.

By the same author
GRAPHOLOGY
LIFE LINES

BIORHYTHMS
YOUR DAILY GUIDE TO ACHIEVING PEAK POTENTIAL

by

PETER WEST

THORSONS PUBLISHERS LIMITED
Wellingborough, Northamptonshire

This edition first published 1984

British Library Cataloguing in Publication Data

West, Peter
   Biorhythms.
   1. Biological rhythms
   I. Title
   612'.022   BF637.B55

   ISBN 0-7225-0967-7

Printed and bound in Great Britain

CONTENTS

INTRODUCTION

From the day you are born there is a continuous range of cycles and rhythms recurring in a never-ending interchange of serial succession within the tiniest cell of your body, in your organic make-up, behavioural attitudes and environment.

The human body is so attuned to this succession of variations that, more often than not, we accept them readily and tend not to recognize them as a cyclic event. The lungs and breathing, the kidneys and their functions, the heart and its beating, day and night, spring, summer, autumn and winter. Some we accept, some we fight, some we do not even know about yet.

Among our bodily rhythms are three clearly-defined cycles that affect our behaviour patterns but which have no cause and effect as such: they are simply continuous, physiological changes. It is these three cycles that this book is primarily concerned with.

The Three Biorhythmic Cycles

Collectively, they are popularly known as biorhythms; individually, as the physical cycle, which has a duration of 23 days; the sensitivity or emotional cycle, with a periodicity of 28 days; and the intellectual cycle of 33 days.

These three cycles control performance in three distinct areas of behaviour. However, it must be stressed that they have no direct cause and effect in themselves but, in each case, are subject to the prevailing conditions of the environment of the moment. It has been found that there is a correlation between the state of the individual rhythm and certain factors in our behaviour. More to the point, awareness of the stage or phase of the rhythm gives the individual an opportunity to correct or adjust his conduct accordingly and the success rate following such action has been phenomenal.

So, these biorhythmic cycles are a potential answer to our on and off days. The intelligent use and awareness of the phase of the rhythms can provide a more positive approach to life and, after a very short while, you will begin to feel far better in yourself.

Biorhythms do not, of course, indicate that on a certain day you will definitely have an accident, throw in your job or win the pools. They do, however, indicate your potential physical ability, emotional sensitivity or mental acuity on any particular day. What you do with this information is, of course, the key to the whole concept of biorhythms. What use, if any, you make of the information that biorhythms provide concerning your capabilities or limitations in each of the three cycles at any given time is entirely a matter of personal choice.

Each cycle begins its individual rhythm on the day you are born and remains constant until the day you die. Because of this regularity, the phase or stage of each rhythm can be computed quite easily for any particular day, whether in the future, the past or the present. All that is needed is your date of birth.

To some people this is the point where a comparison with astrology inevitably arises. But, as yet, there is no discovered relationship, if any, with astrology. Admittedly, both subjects are concerned with the individual and both can advise on a path or attitude to adopt. But astrology advises in more specific terms, though often couched in nebulous phrasing, through transits and progressions, whereas biorhythms can only provide information regarding capabilities and limitations although the method of so doing is quicker. For some individuals, biorhythms appear to be more accurate, but this should not be taken as a comment or slight against astrology and astrologers: the concepts are vastly different.

Nor is there any occult or hidden magic involved. The use of biorhythms is, quite simply, an exercise in calculating a physiological function and relating the results to our behaviour patterns in the light of recent research. Once the results of biorhythmic theories emerged, it was only a question of time before the whole concept became subjected to many other experiments. Not the least of these, and certainly one of the most popular, has been in the field of compatibility studies.

Compatibility

In order to exist, we must get along with others to the best of our abilities. But this is not always very easy. We sometimes find ourselves behaving irrationally towards someone we meet for the first time; something does not quite properly 'click' into place. In effect, our relationship does not get off to a very good start. In other cases, we immediately get along beautifully, as though we had always been friends: there is a noticeable rapport.

Even those we have known for a long time occasionally appear to rub us up the wrong way. At first glance it may seem that our personal biorhythms are not compatible for socializing although, when we check them, we are surprised to find that it is the best possible day of all for a get-together. On other occasions, although we know that we should not be talking to anybody at all, along comes old so-and-so and, somehow, we are lifted out of our depression by the sheer magic of his or her personality. Or are we?

By comparing the biograms of individuals, it is possible to see how we ought to get along with others. It is not infallible by any means, but something like 90 per cent of the time we can see why we cannot get along with him, but we can get along with her. Biorhythms do not have all the answers, but they can, and do, provide an illuminating insight into the whys and wherefores of a relationship.

Your personal biogram reveals your potential behaviour

patterns. It therefore follows that what it does for you, it does for others. So, the logical step is to compare them. The virtually limitless possibilities that this exercise provides in terms of human behavioural patterns and potential compatibility is exhaustive, but exciting.

Another field in which the intelligent use of biorhythms has been gainfully employed is safety, particularly on the factory floor and in all forms of transportation. The success rate in this area has proved almost unbelievably high: masses of statistical evidence exists proving the validity of utilizing biorhythms to reduce accident figures.

The sporting arena has been another success story. Trainers and managers have waxed enthusiastic in the extreme following improved results from their charges once biorhythms have been employed.

In fact, with a little imagination, there are very few areas where the application of biorhythms cannot be successfully employed.

In this introduction I have tried to whet your appetite for what is to come. By the time you have read the book, you will have all the information necessary should you choose to take the subject further and experiment with biorhythms for yourself. I think you will.

1

THE THEORY AND DEVELOPMENT OF BIORHYTHMS

From the dawn of time, all manner of life has responded to the natural cycles that influence all of us, whether from within our bodies or from external sources. The most basic cycle, night and day, was obeyed virtually without question by most animals, man included, either by sleeping in sheltered spots at night if a diurnal creature, or by day if nocturnal.

Gradually we have become aware of other rhythms and cycles, the most natural of which, the seasons, follow on from day and night. Spring, when new life appears in the animal and plant kingdoms; summer, when life flourishes and grows in abundance; autumn, when some life-forms begin their specific annual preparations for death or hibernation from the cold of coming winter.

But not all forms of life respond to this basic periodicity. There are some creatures that flourish during the winter months and, indeed, take on the appearance of a totally different animal during this period. An example of this phenomenon is the stoat and ermine: one and the same creature in reality, but one that manifests differing responses to its natural rhythms according to the season. As the seasons change, so does the outward appearance of this animal.

This change does not occur haphazardly: a definite, traceable pattern exists. The distinctive changes in the ermine's life cycle can be traced accurately by observation; and these observations can, in turn, be checked against statistical evidence recorded over a long period of time.

If this cyclic information is transcribed into graph form, and constantly updated, the changes which occur are much more easily discernible. Eventually, the complete stoat/ermine life cycle will emerge in such a predictable manner that the overall picture can be observed and absorbed by anyone.

Behavioural Patterns in Man

The development of cyclic behaviour in man should not, however, be confused with the natural rhythms of man. Behavioural patterns are those that evolve by virtue of the way in which we live. When recorded, these have produced some startling statistics, as can be illustrated by the following crime information collated in the period between the two World Wars.

Compiled from police files from over 2000 American cities and towns over a period of five years, this information indicated a curious link between seasonal changes and crime patterns. So much so, in fact, that J. Edgar Hoover, one-time director of the F.B.I., was reported as stating that meteorologists could predict rapes as well as storms, to a limited degree, of course.

Crime patterns varied very little from year to year. More murders were committed during the months of July and August than at any other time, particularly during the weekends. Over 60 per cent of these murders took place between 1800 hours and 0600 hours. Burglary, however, was a different proposition. Between 1800 and 0200 hours on a Saturday night in December, January or February was the favourite time.

May saw very little crime, except for an upsurge in dog-bites, more being recorded during this particular month than in any other. However, June was the peak month for suicides, admissions to hospital and marriages; and more cars were reported stolen in February and November than during any other period. The list is endless.

Nearly 3000 years ago the celebrated Greek physician, Hippocrates, noted that we appeared subject to good days and bad. We do not, of course, know whether any serious study was made as to the reason for this, but we do know that this assumption was recorded as being irrespective of whether one were ill or not. It is quite probable that this theory had been expounded earlier but not recorded because it was regarded as of less import than it now is to know these good and bad days.

Hermann Swoboda

Shortly before the end of the last century, Hermann Swoboda, a professor at the University of Vienna, became aware of a slight regularity in certain of man's attitudes. He watched, waited and observed. At long last, among other equally important discoveries, came the realization that there was a definite, rhythmical periodicity which seemingly affected man and his behavioural patterns.

Professor Swoboda continued his research in order to establish whether this fluctuative phenomenon could be predetermined by calculation in some way. Further, he then set out to prove the existence of a 23-day cycle which affected man's physical, behavioural reactions. (Later research has, of course, refined the professor's findings in respect of this cycle.) Coincidentally, Professor Swoboda also discovered a 28-day cycle of emotional reaction and behaviour. This second rhythm was not as easily discernible as the first because it did sometimes coincide with the menstrual cycle in women. But, as the same periodicity was observable in men, Swoboda set to work to establish that a definite pattern existed, irrespective of a woman's natural rhythm. His painstaking research was rewarded by convincing evidence of the rhythms of life.

Primarily a psychologist, Swoboda was naturally analytical and systematic. When finally convinced of his findings, he published his first book, *Periodicity in Man's Life*, followed by *Studies in the Basis of Psychology*. He also devised a crude measuring system and an instruction booklet, *The Critical Days of Man*, to supplement it: the study of biorhythms had been born.

Wilhelm Fliess

But, in a sense, biorhythms had two parents – and neither was aware of the other at the birth. By a strange coincidence, during roughly the same period as Swoboda was conducting his research from a psychological point of view, another doctor was amassing similar information from the standpoint of a practising physician.

Wilhelm Fliess, a Berlin nose and throat specialist, had observed the 23- and 28-day cyclic behavioural patterns for himself. He was the first to realize, or at least to publicly announce, the connections between biorhythms and behavioural patterns. His beliefs were based on the simple theory that each of us inherits both male and female characteristics and that everyone therefore possesses a trace of bisexuality in some way.

Fliess believed that there was a connection between his findings, evolution and life itself, yet his book, *The Course of Life*, went largely unrecognized at the time. Although it was dismissed as too complex and mathematical to understand, this did not prevent Fliess from continuing his researches. He realized the importance of his discoveries and discussed them frequently with another giant of his time, Sigmund Freud. Indeed, Freud was so convinced of the validity of Fliess's work that he utilized his colleague's theories in his own early practice. During the development of Freud's now famous psychoanalytical ideologies, the Fliess theories were frequently referred to and employed.

Early Scepticism

However, these were the early days of biorhythmical research. Like most new ideas and theories, the struggle for acceptance was hard. Even today, we sometimes find it difficult to cope with the newest theories published in some fields of endeavour. It must, therefore, have been far from easy during those early stages of discovery to persuade others to accept the researchers' findings.

In the past hundred years, man's scientific knowledge has advanced significantly. Computer technology is one good example. Thirty years ago, a machine capable of the same functions as the pocket-calculator which most of us now carry around would have occupied a space twelve feet square. And, with the silicon chip revolution now upon us, in what direction will that particular field of technology lead to?

But, during the earlier part of this century, new theories were almost always treated initially with a mixture of prejudice, scorn or suspicion before even minimal acceptance was achieved. Fliess and Swoboda, therefore, had much to contend with. Despite traces of grudging acceptance here, outright repudiation there, they continued to build on their respective theories. Both produced masses of documentation and statistics, and Swoboda's book *The Year of Seven* contained mathematical analyses of the 23- and 28-day rhythmical repetitions displayed by subjects through several generations. This book was vast in concept and represents the foundations on which modern biorhythmics are based.

Alfred Teltscher: Establishment of the Third Cycle

Fliess died in the late 1920s, about the time when the third biorhythmic cycle was recognized. This time, the theorist was not a doctor but an engineer and student of mathematics, Alfred Teltscher. While no real hard evidence stems authoritatively from Teltscher himself, it appears that he established the pattern for the 33-day cycle after investigating the reasons for the variations in students' intellectual capacity. He found that there were predetermined periods when a person possessed poor perception and performance in intellectual pursuits and, equally, periods when the subject could more easily grasp new concepts, perform well and generally exhibit intellectual acuity.

Further to this, Rexford Hersey and Michael John Bennett, both doctors at the Pennsylvania University, conducted similar research and, quite independently of Teltscher, arrived at similar conclusions regarding the 33-day cycle. So, once again, a curious coincidence occurred in the study and discovery of biorhythms.

Early Complexity of Calculations

However, as with the recognition of the two previous rhythms, nothing electric happened and the utilization of biorhythms remained neglected for a long while, only occasionally being referred to in the ensuing years. One of the many reasons for this neglect may have been the seemingly complicated methods of calculation necessary for verifying the stages, or phases of the cycles. None of the original discoverers seem to have been able to simplify the methodology in a way which would have proved acceptable to the layman and professional alike. Yet, in practice, the calculations are very simple.

We merely calculate the number of days that the subject has been alive; counting the birthdate as day one, all that is needed is the total number of days from the date of birth up to (and including) the day in question. This total is then divided by the number of days in the required cycle; that is, 23 days for the physical, 28 days for the sensitivity and 33 days for the intellectual. The remainder figure resulting from these divisions evaluates the stage of the individual rhythm; if there is no remainder, the day in question is the start of the new cycle.

Yet there appears to have been a stumbling-block or mystery attached to this area of biorhythmic study. Admittedly, some clumsy attempts at a satisfactory formula were made: slide-rules were produced by various individuals; sets of complicated tables were published by those interested in biorhythmic theory or simply as a result of the mathematical challenge presented by the problem. Whatever the reason, public interest in the subject declined. With the exception of a few dedicated individuals, the whole thing just seemed to lurk in the pending file until, suddenly, in 1939 interest was again aroused as the result of a new publication.

Hans Schwing

Hans Schwing, of the Swiss Federal Institute of Technology in Zurich, produced a 78-page treatise in a comparative study of accidents and accidental death statistics. This publication did not lead to a revival of interest in biorhythms as such, but it did emphasize the public's potential interest. Over the next twenty years or so came a spate of similar dissertations. These provided many interesting advances in the theories and proved that the time of birth might be valuable in assessing biorhythmic phasing. However, once again, the inevitable comparisons with astrology arose.

This particular period was one of strong interest in astrology, again by a few dedicated adherents who were much more interested in the validity of a system rather than in the end product. Until very recent years, astrology had not been regarded as a 'respected' science or art (call it what you will); nor did biorhythms attract the necessary research it really warranted. Again, it must be stressed that there is still no real correlation between the two studies, except that we now accept the invaluable proof of periodic or cyclic behavioural patterns that affect us all, whether regarded from the astrological or biorhythmic standpoint.

New Interest in Biorhythms

Low-key public interest in biorhythmic studies was once again displayed, occasionally interrupted by the odd publication or two, although there appears little reason for the sudden explosion of popularity that the subject enjoys today. Now, there is hardly a country in the world which does not have at least one biorhythm society or association. In the last very few years, literally, public interest in Britain has developed enormously, slowly at first, until the current tremendous boom in popularity. We, in Britain, have not quite reached the stage that exists in America perhaps, where one can obtain a weekly forecast from a slot-machine for a few cents, but we do have several respected and well-conducted researchers and publications on the subject. Occasional publicity in newspapers has helped to generate further interest and one magazine in particular offers a readers' biorhythm service for a very reasonable fee with remarkable response and requests for repeat orders a regular feature.

So, at the present time, we have an established and respected study and practice which, if intelligently employed, can and does lead to very significant improvements in the way in which we live. How, then, do these three rhythms affect us, how are they determined and what exactly are they?

The Phases

It is of paramount importance to realize that the three cycles of biorhythms, irrespective of their phase, do not have a cause and effect in themselves. Fundamentally, they are continuous, physiological changes and awareness of them can help you to plan your way of life much more effectively. Because of the phasing of the cycles, you will either tend to perform well or give a less than average attendance to matters of the moment, subject to prevailing conditions.

Each rhythm begins on the day you are born and continues its individual course throughout life, only ceasing at death. Everyone has them and is 'subject' to their influence, although a tiny percentage of the population does not 'conform' to the established patterns all the time.

The first half of each cycle is the plus, ascending, developing and progressive period. Confident and aggressive, full of vim and vigour, mental perception at a peak, you will perform well until reaching the zenith of your powers midway through this first phase. Your capabilities will remain at a high level, then gradually tail off until the rhythm reverts to the second half of its cycle.

This second phase is the rejuvenation period, as though it were a recuperative period after an operation. This half of the cycle sinks to a nadir, again at the midpoint, then begins to steadily progress towards the positive phase once more until the cycle has been completed. This pattern is repeated, continuously, throughout life.

Critical Days

The days on which the cycles begin or transit from one phase to the other are known as 'critical' days. Publicity of these critical days, or perhaps merely the label 'critical' which has been applied to them, seems to have served to popularize biorhythms more than any other aspect, for it has been proved, statistically, that more accidents occur on critical days than at any other point in the cycles. The probability of accidents occurring on these days is very high, whether through lowered physical vitality, irrational emotional behaviour or inferior mental perception causing the subject to be more error prone.

Each rhythm has three critical days: at its beginning, the commencement of the cycle; at the half-way stage when it changes from the positive to the negative phase; and at the end of the cycle, which, of course, is also the beginning of the next positive phase. (There are also other significant points in each of the cycles, but these are detailed in the chapters which deal with the individual rhythms.) The designation 'critical' is somewhat of a misnomer, however, because nothing critical does occur, save for the actual changes in the cycles.

In most aspects of life, transition from one phase to another may be termed a turning point, a psychological moment, favourable or providential; or, perhaps, disturbing, unlucky, inauspicious or unsuitable, depending on individual circumstances. Understanding that these various meanings can be applied to the 'critical' days in your biorhythms will enable you to utilize them to advantage.

A change of job may, for example, be favourable in the greater analysis, but it is also disturbing. When a child leaves junior school to attend senior school it is a turning point in his or her academic career, but it may also prove unsuitable because of the environmental changes involved. A new manager may be appointed in charge of your department, shop or factory at a psychological moment either favourable or unfavourable to you. Any change in a regular facet of life may, therefore, have an effect that can be determined in some cases, but only guessed at in others.

Abrupt changes are a different matter altogether, however. Let us assume that you return home each day in exactly the same manner whatever the circumstances and irrespective of weather conditions, season or whatever. One day your regular train is cancelled, something which has never occurred before, or there is a party of tourists occupying the carriage you normally enter. Your immediate reaction will naturally be self-defensive, but it can manifest in different ways according to the individual personality. Despite the way in which you react in such a situation, you will be temporarily off-balance.

Perhaps you have arranged to meet someone at a specific time and place, with no doubts that the details of the meeting are correct. You arrive a few minutes early but, after twenty minutes or so, realize that the other party is not going to arrive. Your reaction is one of temporary emotional disturbance and perhaps intellectual annoyance at suddenly being faced with unexpected alternatives to rectify an unfamiliar situation.

Maybe you have planned and worked hard to produce a sales programme which, you believe, cannot fail because of the untold hours you have spent verifying all the details. But your employer rejects your scheme out of hand, either qualifying his decision or perhaps not even bothering to give you a reason at all. Your inner reaction will leave you temporarily off-balance.

Such everyday situations, and your reactions to them, are what critical days are concerned with. You are temporarily off-balance during these transitional periods which can last for between 24 and 48 hours: a long time to be off-balance, even to a minor degree. Suppose that one of the examples quoted above occurred while your biorhythms were in one of their transitional periods, you were experiencing a critical day, a double critical day or, even worse, a triple critical day. In such an event, your normal code of behaviour could be swept aside in a moment of emotional irrationality, intellectual blindness or physical rage: you would be disturbed.

The Grand Triple Critical Day

There is one point in everybody's life where all three biorhythms reach the same phase and stage as on the day of birth, to restart their respective cycles all over again in exactly the same patterns. This 'grand triple critical day' occurs 21,252 days or 58·2 years (58 years and 66 or 67 days if an extra leap year is involved) from the date of birth. During this period, 924 physical cycles, 759 sensitivity cycles and 644 intellectual cycles have occurred. (The figure 21,252 is arrived at by multiplying the number of days in each cycle together: 23 × 28 × 33.)

Up until this point, each rhythm has had a succession of critical days, sometimes coinciding with the other rhythms going from positive to negative phase at the same time, resulting in double or triple critical days: but without all three cycles reaching the identical point as on the day of birth. There are at least six critical days each month, sometimes

there are eight, in everyone's biorhythms. Allowing for the
basic six occurring in an average month of thirty days, this
means that you are in a critical phase of your biorhythms for
20 per cent of your life, every month.

Emotional and Physical Double Critical Days

During the grand biorhythmic span of 58·2 years, there are
numerous occasions when a double critical day occurs in the
emotional and physical rhythms, and it is this particular
combination which is statistically proven to be the most
accident-prone period. Single critical days can portend
problems of their own, of course, because you are likely to be
temporarily off-balance on these days, but the emot-
ional/physical double critical days should be treated with
caution; awareness of the inherent problems of these days may
help prevent distress or accidents. Simply checking back to a
date when something went wrong has a more than 60 per cent
probability of revealing that you were in a critical phase on that
day. More to the point, you may even recall saying at the time
that had you been 'aware' of certain information, the incident
may not have occurred.

Accuracy of the Birth Date

In the physical rhythm, the critical days are on days 1 and 12;
in the emotional cycle, on days 1 and 15; and intellectually, on
days 1 and 17. There is a strong possibility that the dates you
are investigating may fall on either side of these critical days,
but there could be a simple explanation for this. Your time of
birth may have been very early in the morning and, while
many doubt the validity of the actual birth time being
important in this respect, others regard it as vital. There is
certainly some evidence that, where the timing of a critical day
seems to be out by 24 hours, the time of birth is the best factor
to consider. Similarly, a very late time of birth could have an
effect in the opposite direction.

Alternatively, of course, it may be possible that you have the
wrong date for your birthday. It may seem strange in this
modern world with all its advanced technology to make such a
statement, but it can and does occur too frequently for the
analyst to ignore the possibility. At this point, it would be wise

to remember that biorhythms are not the be all and end all of
accuracy, errors can occur but, statistically, the chances are
strong that they will not.

Using Biorhythms for Planning Ahead

Now, let us consider how biorhythms can be utilized for
planning ahead. If you have important dates or events coming
up in the future and wish to know your capabilities it is a
simple matter to check the particular phasing of your
biorhythms for the day in question and plan accordingly. If
the relevant cycle appears unfavourable, allow for the
possibility of error in your behavioural patterns or, if you can
change the proposed date to a more favourable time, do so: a
simple adjustment is all that is needed. Once again, the
inevitable comparison with astrology may occur, but you
should disregard this as there is no correlation yet proven.

Despite all the advances which have been made in
psychological analyses of behavioural patterns, an element of
'chance' can always enter the equation. However, to give some
idea of the valuable potential of biorhythms and their use, let
us return to the thesis of Dr Schwing, published in 1939. This
report, an extremely accurate and precise analysis, was based
on 700 accident cases, with a further 300 cases of deaths
recorded in the city of Zurich's archives.

Schwing set out to prove the validity of biorhythmic theory
and its relationship to a detectable pattern, or cycle of life.
Using the complete biorhythmic span of 21,252 (23 x 28 x 33)
as a basis, Schwing's calculations showed that there must be
4327 days on which one or the other biorhythms must be at a
critical point, with the remaining 16,925 days being
comprised of mixed rhythms. In percentage terms this was
expressed as a ratio of 79·6:20·4 per cent.

Schwing demonstrated that 322 accidents were recorded on
single critical days, 74 on double critical, and 5 on triple
critical ones. The remaining 299 accidents occurred on mixed
rhythm or normal days. Thus, nearly 60 per cent of accidents
(401) fell on critical days: representing 20 per cent of the time,
whereas only 40 per cent (299) fell on the remaining 80 per cent
normal, non-critical days. Food for thought indeed.

Another report, published in 1954, resulted from the study

of 497 accidents involving agricultural machinery. The author of this report, Rheinhold Bochow of the Humbold University of Berlin, found that only 2·2 per cent of the accidents fell on normal, or mixed rhythm days. However, 26·6 per cent fell on single critical days; 46·5 per cent fell on double critical days; and 24·75 per cent fell on triple critical days. The astonishingly high percentages shown in this particular series of results spurred other researchers to further study.

Application of Statistical Evidence

If the results of these accident statistics were to be used as the basis for future analyses by, say, a small transport company, surely the possibility of accidents could be reduced? Drivers could be taken off the road on their most accident-prone days and be given alternative work, or they could be taught the fundamentals of biorhythmic theory in order to become more aware of the potential hazards of particular days in their cycles. A good company, with good employer/employee relationships, could even work out a rest-day rota system so that no real hardship would be involved either in earnings or in the loss, however temporary, of personal prestige. This is exactly the sort of scheme which has been implemented by some companies, all over the world.

All kinds of business concerns, not only those directly concerned with transportation, have utilized biorhythmic studies in respect of their staff, and the success rates achieved speak for themselves. Accident figures are down, the insurance companies are more than happy as a result, management are delighted and the personnel involved feel the better for it. All this had, naturally, led to increased productivity too: an additional bonus.

Other Uses of Biorhythms

Biorhythms can also be used effectively to check your compatibility with others. We all want to get along as well as possible with other people, but this is not always easy to accomplish. However, we might as well make the best of what we can. So, if there is a way of improving relationships, then the obvious step is to use whatever mode of improvement is available. Your personal biograms reveal your potential behaviour patterns. It therefore follows that if they do so for you, they do so for others also. The logical next step, then, is to compare them. And, in the field of biorhythmic compatibility studies, an astonishing success rate has been recorded once the basic principles have been understood and simple rules observed.

The use of biorhythms does not stop here, however. They can also be used for diet courses, giving up smoking, improving sexual harmony, sporting successes, academic studies, effective holiday planning, redecorating, gardening ... the list is endless. Remember, once you have your biorhythms charted, you can live a far more positive life in every sphere – provided you work at it.

2

THE PHYSICAL CYCLE

This rhythm is totally concerned with all the physical
phenomena of performance and capability. The range of
possibilities and potential is enormous and, when taken into
consideration with the other two rhythms, it still rates high.
This is because whatever activity may be under review, the
physical ability to carry through the tasks at hand must be
favourable for positive results to be achieved.

Sometimes the effects of this cycle are hardly noticeable, at
others they are painfully apparent, occasionally surprisingly
so in the course of normal daily routine. It does not matter
which day we choose, so let us start from the moment when we
hear the alarm in the early hours. Your arm snakes out and
shuts off the hideous clatter, or it gropes blindly and dully.
You leap out of bed, or you stagger.

The Positive Phase
If you are in the positive phase of this cycle, you should begin
to come awake feeling good and the men may be particularly
aware of this. Shaving proves reasonably easy, irrespective of
the method employed: the skin feels good, smooth and elastic.

The blade will glide over the face with far less chance of a nick and the skin's response to an aftershave astringent sets the tone of the day.

There is more chance of feeling like a light meal, at least when this cycle is in the plus phase, and, depending on normal personal habits, what you eat will be well prepared: no burnt slices of toast today.

The usual walk to the station scarcely has an effect. You notice that you pass others who, on other days, have passed you. You will not object to standing on the crowded train or sitting in a non-smoker even if you are a heavy smoker yourself. The walk from the station to the office, factory or workplace is taken – literally – in your stride. If it is an office, you will experience a certain amount of restlessness and will not want to remain seated at your desk for too long at a time: the slightest excuse to get up and walk around will be taken.

You will feel almost frisky, but with a slight tendency towards irritability without knowing why. In fact, the cause will simply be a lack of exercise: those short excursions you feel like making represent your need to work off an excess of physical well-being. The manual worker will achieve all his tasks confidently and sometimes ahead of schedule which, if on an assembly-line, could make him feel slightly bored, again leading to a touch of irritability. The lunch break will be relished more for the opportunity it provides to get some physical exercise rather than for eating.

The journey home will be approached with even more zeal than usual and, on reaching home, the slightest excuse to walk the dog, mow the lawn, redecorate or put some time in on the car will prove welcome. Better still, an evening spent playing tennis, football or any other sport is just what the doctor ordered when in this positive phase of the physical cycle with its attendant feelings of physical well-being.

The Negative Phase

But, in the negative phase the story is reversed, of course. Getting up will be a struggle, staying up a bigger one still. For the men, shaving can be a hazardous affair because the skin will feel rough and wet shaving may produce a cut which, even if it is only a slight nick, will feel quite painful. Breakfast may be off, you cannot bring yourself to eat a thing and may only just be able to swallow one cup of tea or coffee. The journey to the station seems too long: it could even actually take an extra five minutes the way you feel. Standing would be painful and you will have to stand even if there are seats in a non-smoker because you simply must have a cigarette to keep your flagging spirits going. The escalator or lift will be preferred to the stairs and, if it is possible, you will wait for a bus rather than walk the rest of the way.

On reaching the office or elsewhere, you will sink gratefully into the nearest chair for, by now, all you want to do is die. Work loads, no matter how light, will take their toll. At lunch time, even if you have been glued to your chair all morning, you will just want to rest. A few sandwiches will suffice and you may even put your head down for a short nap. Somehow you will struggle through the rest of the day, but come the time to go home you may even feel like staying the night just so you will not have to spend all that energy again!

The journey home is accomplished, however. Once there, all you will want to do is sit down in front of the television. The dog can walk himself, the grass can grow, the decorating can wait and the car can have its service another time. At the end of the evening, getting to sleep may seem a little more difficult than usual but, once achieved and although deep, it may not feel as though you have had adequate by the time you wake up the following day.

Now these two comparisons may seem a little dramatic, but if you care to spend a short while thinking about it you will remember having days like this yourself, perhaps not quite so exaggerated, but we all experience good days and bad. A systematic check will show we have these in cyclic form and that they run more or less along the lines described.

Just occasionally, certain days may stand out in the mind. Maybe a day when you fell for no reason; seemed to have pulled a muscle carrying the shopping; slipped off the ladder while decorating; or an incident involving some form of physical phenomenon. This may have occurred while you were having a critical day and, unconsciously, falsely misinterpreted your capability. You may have agreed to go kite-flying with the children and overdone the exercise altogether in the excitement of the endeavour. It may sound strange or even funny, but this is exactly what biorhythms are all about: determining your capabilities for any given day.

Reading the Biogram

If you look at Figure 1, you will see that a biogram for one month of thirty days has been prepared showing the duration of the physical rhythm which, in this example, starts on the first of the month. The 'O' line merely indicates the norm but, as the cycle is continuous, there is no actual norm: the graph line either crosses or is above or below this base line. In fact, the point at which the curve crosses this base line is the critical day. So, straightaway we can see that there are going to be three critical days this month, two when the cycle moves into the positive phase and one when it goes into the negative stage: on the 1st, 24th and 12th respectively.

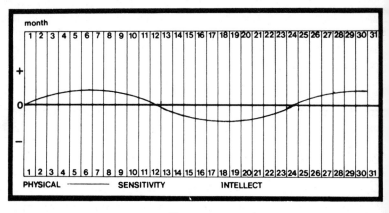

Figure 1

In practice, these biogram charts show all three rhythms together, either in different colours or style, so that the exact stage of each rhythm may be compared against the others. In our example, the physical cycle is in the 7th day of its new positive phase on the 30th of the month.

Effects of the Positive Phase

The cycle starts on the first day of the month and gradually rises into the positive phase, achieving its maximum extent at day seven: the 'mini-critical' day. Your physical abilities are at maximum potential on this seventh day peak and, if you are undertaking any form of special physical activity, you ought to be able to give your best performance. This is the stage of the positive phase where you should perform superbly well physically, although you may just possibly overdo things and cause physical injury or discomfort in some way.

There are many examples of the validity of the physical cycle's positive phase, but perhaps the most impressive is that of Mark Spitz who won no less than seven gold medals for swimming in a very short space of time at the 1972 Olympics (see p.99). Between 27 August, a rising critical day, through to the next critical on 7 September, with the mini-critical occurring on the 2nd, Spitz achieved this unprecedented and unequalled world record. (Although this chapter is only concerned with the physical rhythm, it is worth noticing that, at the time, his emotional cycle was also in the plus phase, peaking to its mini-critical day on 1 September.*)

Effects of the Negative Phase

In the negative phase, the lowest point of the cycle is reached on day eighteen. It is at this stage that the rhythm begins to turn once again into the positive phase which it enters on the 24th day. This is also the most negative part of the cycle, the trough, when the mini-critical day occurs: the day of maximum lethargy, no energy and no inclination for anything.

Athletics on this day would not be recommended save for routine practice. Often it is on this negative mini-critical day that people over-reach their potential by trying to do far more than they are capable of doing. Finishing the decorating or gardening, trying to complete outstanding tasks that for some

* See Figure 32, Chapter 7.

reason never seem to get done on schedule but have to be for a variety of reasons: these are just the sort of tasks which are unsuccessfully attempted at this stage. For, if there is going to be an incident or accident caused by or through physical incapacity, this is the day of highest probability.

Making Necessary Allowances

Again, it must be stressed that biorhythms do not have a cause or effect in themselves, but are merely guidelines for timing events for the best results. There is no reason why you cannot do all sorts of physically strenuous jobs in the negative stage, or on critical or mini-critical days, *provided that you allow for the fact that you are in these particular conditions of the cycle.* After all, sportsmen, especially the professionals, have to compete continually: it is their livelihood. But most of them know their capabilities backwards and are aware of cyclic behaviour patterns in their make-up. They allow for their temporary weakness and capitalize when they are on form.

In football, for example, it is frequently stated that the centre-forward of a particular team is 'off form' because he misses easy goals or other opportunities where he would ordinarily move in with his accustomed skills and win the moment. Then there is the cricketer who consistently bowls erratically for a few days, or bats in an appalling manner where he would normally have his hundred up; or the snooker player whose 'eye' is out where he normally has been known to clear the table. All these are examples of below average physical performance in the negative phase of the physical cycle.

It works in reverse, too. The champion may be off form and the outsider comes in from nowhere and beats him. Both are professionals, both are good, but the champion usually has the edge because of other special skills. Yet, if his rhythms are in the wrong phase when his opponent's are in the right one, the advantage will lie with his opponent: that is how new champions are made. At least it is an explanation of how a man may temporarily achieve the fame he seeks by eclipsing the proven man at a fortuitous time but is unable to maintain the new-found position.

Applying the Biogram

Simply by charting your personal rhythms and noting their phasing, you can, with simple adjustments, achieve better results which must inevitably lead to a better way of life. The theory and practice of biorhythms can be applied in all aspects of life. In early spring the garden begins to beckon with its annual need to be turned over and generally tidied up; at the same time, the house would probably be transformed by an attack of spring-cleaning. Looking at Figure 1 again, it is obvious that the best time for both jobs would be between the 2nd and the 11th or between the 25th and 30th. However, there is no reason why you cannot attempt the tasks on the other dates, that is between the 13th and the 23rd, provided you realize you will not have the same capabilities as in the first two periods. Allow for this, do not overdo it and jobs will become easier to take on and finalize.

Physiological Basis of the Physical Rhythm

In all probability, the physical rhythm relates to our basic muscle fibres. As everything in our bodies has a rhythm or definite cyclic existence, in some way it may be that everything to do with our physical capabilities culminates in this expression of the physical rhythm. We all have biological clocks and we all respond to them.

Some people vary slightly from the norm: they appear to be out of step with the rest of their immediate circle. This may be observed in the phenomenon of 'day' people and 'night' people or 'morning' and 'evening' people. Their personal clocks are the same as everyone else's, but they appear brighter or physically more active at different times of the day. Every individual responds to his personal 'inner' clock in exactly the same manner, but the timing may differ.

In many cases this may be traced to the time of birth. Often, people born after noon are 'night' people, coming alive around 10 or 11 o'clock in the evening and staying quite bright through to the early hours. Those born in the early hours rarely like the early hours in practice. This is not infallible and there are many exceptions to the rule, but it could account for many personal foibles.

It is no good playing a round of golf if you start by feeling tired and finish exhausted and irritable. It would be sensible to suggest to your partner that nine holes would be better because you are feeling under par. He might be more amenable in the circumstances; knowing that you are at least prepared to make an effort, even if he does not know why. This is half the battle of successful relationships and improving compatibility with those around you. Compromise, a dirty word to some in this day and age, is effective when used in a judicious manner. Both parties want to enjoy themselves, so try to arrange the maximum pleasure possible in the right atmosphere at the right time: simple adjustment is far better than an outright refusal.

The Health Factor

In the negative phase of the physical cycle a strong health factor is involved. You are then far more prone to catching chills and colds, getting indigestion because of inadequate or irregular meals and upsetting normal regular personal body functions as a result. During this phase you are more susceptible to pain, bleed more freely in the event of injury and recover more slowly than usual from illnesses. In fact, so much so, that some doctors time operations to coincide with the plus phase of your physical rhythm in order to facilitate recovery. Pain is then easier to cope with and your body is in better fettle all round. Post-operative shock to the body is a serious business and can kill, but the chances are considerably lessened if such time factors are taken into account.

Some patients on supervised courses of drugs may find that they are strictly informed of dosage and dosage times these days. It has been found that the old idea of 'three times a day after meals' can be very dangerous in certain circumstances. Doctors now know that what is more or less acceptable at 8 o'clock in the morning can be positively lethal at 4 o'clock in the afternoon.

Power Surges

We all recall those occasions when we felt out of sorts at about four or five in the afternoon: dreading the idea of having to meet a social obligation of long standing later in the evening and wondering how we are going to cope. Later in the evening, we suddenly realized that we, who were half dead some three or four hours ago, are now on top of the world: we simply do not feel the same person. There are several possible reasons for this. The normal pulse variations during the 24-hour period may be responsible. If the average rate is 72 per minute, in the evening it may be higher or lower by as much as ten beats, making a considerable difference in our physical responses. Most of us are at our lowest ebb in the early hours: even if we are 'night' people, we bear little resemblance to the active chap of some twelve hours later when we are virtually at the opposing peak.

Another reason for a dramatic change in our responses could be the actual twelve hours or so when we experience the critical day change period. It has been reported by some people that they can feel the surge of power, albeit brief, when undergoing a physical critical day moving from negative to positive, and, when switching the other way, a point in time when they experience a short period of extreme fatigue, lethargy and temporary intolerance. Others have marked a critical day in their diary, often without knowing a thing about biorhythms, as the day when they suffered a bad bout of indigestion brought about by poor meals or the timing of meals. They perhaps ate the wrong things, making do with a stodgy sandwich when the body was crying out for a square meal.

This is a major fault with many long-distance drivers who tend not to take adequate rest away from the wheel and who prefer to get the job or distance done rather than take time out for a meal. Even without a knowledge of biorhythms it makes sense to have a good meal at regular intervals so that the body can maintain a steady stamina flow. Failure to do so, especially when driving regularly, is often the first step toward a potential accident.

Taking Care in the Negative Phase

If normal stamina is low on a physical critical day, a meal forgotten or quickly glossed over, the mental responses will become dulled. In the event of quick reactions becoming necessary, only a half-hearted response is likely: this can be the difference between life and death. Excessive smoking when not eating regularly also tends to dull the normal reflexes. As a substitute for a meal, heavy smoking often causes headaches and, if these occur during the negative phase, the body has no reserves to fall back on. It is like hitting a man when he is down, only in this case carelessness may kill.

Equally worthy of attention on such days are drinking habits, and not just alcohol. The body requires minimum and maximum amounts of liquid, varying according to individual need. A lack of care in all dietary matters will affect sleeping habits, and if you fail to take sufficient care in the negative phase of the physical rhythm you are in effect punishing your body even more.

None of us is perfect and it is not always possible to take the rest and diet we should.Occasional lapses will not hurt a lot, although they may create unnecessary problems for some. Whenever possible, if you can maintain a correct attitude, especially at the critical times, you will help your body to function properly. Regular meals, moderate exercise and the establishment of routine habits to conform with your biorhythmic patterns will increase your capabilities and achievements.

Those of you who suffer from rheumatism, arthritis or asthma should find it easier to avoid crippling attacks by keeping a discerning eye on your physical rhythm, especially if adverse or uncertain weather conditions are forecast. People with high blood-pressure or heart conditions should avoid driving when their biorhythms are in an adverse phase. In these particular conditions, never drive on an empty stomach because the resultant lowering of blood sugar can lead to accident-prone situations.

Reading the Signs

Knowing that your potential is low alleviates some of the tensions otherwise encountered and lessens the probability of problems arising. But, if they do arise, being able to read the danger signs and make the necessary adjustments in plenty of time becomes easier. Driving, especially, may become hazardous on critical days: even if you are not initially the cause of an accident. Remember, negative to positive days are just as conducive to such situations as are the positive to negative: your reactions may be imperfect. Those who sense that little surge of power on the negative to positive changeover days speak of a tremendous feeling of well-being and physical bonhomie. In fact, they may be lulled into a false sense of ability and their judgement become faulty as a result.

Similarly, long spells at the wheel on the reverse critical day can also take its toll because you fail to notice just how much energy has been drained from your body. On these days the wearing of a seat belt helps combat fatigue up to a point. On long-distance journeys it is surprising just how much the body moves when seated freely. With a belt restricting the body movement considerably, energy is conserved that would otherwise be frittered away in a non-productive manner.

Educating the mind to the capabilities of the body is reasonably easy for most people. Despite our occasional lazy streaks, nearly all of us will sometimes overdo things if out of the ordinary tasks need to be performed.

These are just the occasions when accidents are most likely to occur. We become so caught up with the matter at hand that little things pass us by. We tend not to notice or choose to ignore little aches and pains, but simply put them down to unaccustomed physical activity. We are right: but we forget that these are part of the body's natural early-warning system. Swimming is a good example: it is a pastime that really does take far more out of us than we fully appreciate. On critical days, we should be very careful indeed; not just because cramp may set in and temporarily cripple us, but because of the blind panic that often drowns the victim. The two are inextricably linked: if we had not overdone the one, we would not have experienced the other extreme. More cases of drowning occur on double critical days, that is, physical/ emotional or physical/intellectual critical days, than at any other time. With a little thought, a lot more of us would still be alive today.

Overstepping the Limit

Many of us live on nervous energy, dismissing the physical limits we think we may easily ignore. But, sooner or later, there comes a time of reckoning. Careful monitoring of the physical cycle can alleviate a lot of the unnecessary stress which we could well do without. These are stressful times that we live in, and stress is a killer. In simple terms, stress may be caused by merely failing to realise how far we are driving ourselves (or being driven) beyond limits we do not recognize.

Coming to terms with this problem is not always easy: our present mode of living, particularly in cities, naturally involves a certain amount of stress. The bus is cancelled or late and, when it does eventually arrive, you cannot get on because it is full. The boss asks too much of you, but you will not or cannot say no. Friends ask for help when you really do not feel like giving it, but you do. You ask someone to do a favour and they refuse, resulting in a little more stress than you have bargained for. These are the sort of occasions when your physical welfare should be carefully monitored. In the negative phase, try to ease off as much as you can; in the positive stage, do not be tempted to overdo things.

Biorhythms and Slimming

Slimming is a good example of what I mean. Most of us get on the scales at the wrong time of the day or the wrong time of the year and wonder where all those extra pounds have come from. Immediate decisions are made to go on a crash diet: no bread, no potatoes, no more sticky buns, ease off the drink and take more exercise. The results are dramatic and immediate in most cases: one minute the body is being abused and over-indulged, the next it is hardly being cared for at all! In the wrong stage of your physical rhythm such foolhardy action can cause havoc in every sense of the word. Unfortunately, it is also doomed to failure despite any willpower being shown by the individual concerned. Instead of attempting something that you suspect cannot succeed, utilize your biorhythms intelligently so that a diet is started when it will have more than a 50 per cent chance of success. Remember: willpower must be a major element, but you can boost it when it starts to flag if you time it right.

Start your diet, a sensible, well-planned affair, when the physical rhythm is in the negative phase and preferably just after an intellectual rising critical day. You need to keep willpower under control, and this will be easier if the intellectual rhythm is in a positive phase. Also, it will be easier at this stage because your body will require less food. On the next physical critical day your body will want a little extra sustenance, but if you are careful to balance the extra intake with a little additional exercise, judicial balancing will outweigh any negative feelings. Extra portions of low calorie foods will not hurt if you keep them under control.

If your emotional cycle is in a positive phase you will feel 'right' and have the necessary emotional approach to your dietary plans. But, if it is in a negative stage, you may feel slightly low and even a little sorry for yourself. This should present no difficulty, however, as you attack what is a mainly physical condition when in the right physical phase. The nearer you start your diet to a critical day in the physical rhythm, the longer you have to establish the routine: about ten days. The correct business-like approach adopted at this stage and strictly adhered to will have far more effect than a haphazard attempt.

Naturally, much depends on how overweight you are, but an average person could easily lose half a stone in twenty-eight days without ill-effects.

There are numerous ways in which your physical rhythm alone can be utilized to advantage. Remember, you do not have to stop doing things just because your rhythm is in its negative phase for about ten days, or is at a critical point; all you need do is ease off the pressure to compensate the body for its lack of energy or over-abundance. Once you enter into the swing of things and start to enjoy the new routine it will soon become second nature to utilize your biorhythms intelligently. Prepare for the negative phases and take life at a steadier pace while they last; make extra special efforts on the critical and mini-critical days; during the positive periods make the most of your capabilities and capitalize all the way, but sensibly.

There will be some occasions when you cannot make the simple adjustments that you would like, but at least you will know just how far you can safely go without incurring injury. That, in itself, must be reward enough.

3

THE EMOTIONAL CYCLE

The 28-day emotional cycle, sometimes referred to as the sensitivity cycle, is concerned mainly with mood, sensitivity and social ability. It is the easiest rhythm to chart, but probably the most difficult of the three to contend with and, for a variety of reasons, is also the most misunderstood.

Because of its duration, which coincides with lunar phasing in astrology, it is often the key to the comparisons made between biorhythms and astrology. The correlation, though, is a vague one; nor is there any relationship between this rhythm and the female menstruation cycle, even if the two may coincide for a while, because this particular cycle never varies, whereas the female rhythm does.

Emotional Critical Days
Like the physical cycle, the emotional rhythm starts with a critical day, has its transition day from positive to negative fourteen days later and ends on the third critical day fourteen days after that. It, too, has its mini-critical days on day eight in the positive phase and at day twenty-two in the negative stage. Because of this phasing, those born on a Monday will

always have alternate Mondays marking peaks or troughs in their emotional cycle, with the attendant behavioural patterns associated with them, while every other Monday will be a critical day. This could account for the 'favourite day' theory. Not everyone will find this theory infallible, but it is quite surprising how much of a guide it can prove when trying to see which were the good days and which were bad.

On an emotional critical day, you are at your most susceptible to a variation of emotional reactions, irrespective of the direction of the rhythm at the time. Irritability, insensitivity or irrationality can be the response shown to almost any situation. Differing environmental situations will not make much difference at such times although, in normal circumstances, environment and prevailing conditions are taken into account before determining response. Even the calmest of people have been known to explode at the critical psychological moment; conversely, the most volatile have been known to pass through this period without raising an eyebrow.

As this rhythm controls everything to do with emotional response at all levels it takes on added importance in relationships, potential accident situations or anything requiring drive and enthusiasm. There is little point in giving a party if it coincides with day twenty-two of the emotional cycle, a physical negative stage and an intellectual critical day, as even the most cynical of critics should appreciate. However, if such an adverse situation is noted and its limitations recognized, such a social event could still take place as long as the individual makes a serious attempt at remaining on an even keel.

The Emotional Positive Phase

Governed largely by the nervous system, this cycle was described by Fliess as the manifestation of the cells influencing the feminine inheritance in our make-up. The first half of the emotional rhythm relates to the plus or positive phase when we are more cheerful, responsive and optimistic. This period is favourable for all creative enterprises, romance, friendship and co-operation generally. Co-ordination plays an active part in this cycle, too, for the nervous system has to 'feel' that things are right before this will be manifest. In fact, co-ordination is markedly absent during the negative phase.

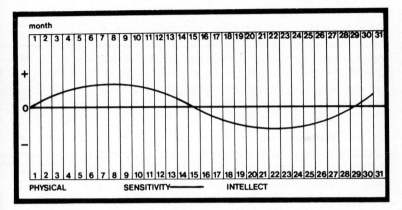

Figure 2

By the eighth day of the positive phase, the individual's sense of well-being and sociability will be well defined. This is the point of peak performance in emotional sensitivity. However, on this mini-critical day, there is the risk of becoming over-confident and overdoing things because of the way you feel. After this stage, the rhythm curves down towards the critical day halfway through the cycle, and your performance, while still good, begins to lose its edge.

The Critical Day
On the critical day itself, almost anything can happen if any kind of stress is generated through emotive issues. Of the three biorhythms, the emotional is the one most prone to error and accident. Your thinking is coloured by your emotional reactions to what is going on around you. You may start the day well but, if the slightest thing upsets you, it is possible for World War III to begin: you could become ultra-touchy and over-sensitive to an immoderate degree without realizing it.

If you are driving and you are overtaken by another vehicle, you could chase after it just in order to overtake the other driver in turn. It sounds silly and dramatic, but there are thousands of such foolhardy reactions recorded for emotionally critical days. You may have planned to go out for the evening and have been looking forward to it all day but, if your partner delays or criticizes you in some way, no matter how well-meaning that criticism is, you are liable to over-react. You may sulk, refuse to go or start a row quite unintentionally. At work, if something goes wrong with the normal office routine, you could become easily exasperated: ranting and raving may put you outside the front door permanently. Or, perhaps, you use your authority and dismiss a subordinate for a mere triviality which, in your present mood, appears to be a heinous offence.

The Negative Phase

As this cycle switches to the negative stage you become less co-operative generally. You become moody and easily depressed, touchy and sensitive to issues that ordinarily do not bother you unduly. You feel that the world suddenly owes you a living and that everyone is against you. The most marked effect is felt by those who have to work with you, or your loved ones.

However, there is another manifestation to this negative phase of the emotional cycle. You become a demon-shopper, finding fault with everything you want to buy and criticizing everyone and everything from the poor shop-assistant to the decor of the shop. You may check your change so deliberately that you insult the intelligence of the cashier and, in extreme cases, it is even possible to become barred from certain stores as a result of extremist behaviour.

Oddly enough, you may be unaware of your extreme behaviour. On day twenty-two, the nadir of the cycle, everything and everyone is your enemy: you have a totally negative outlook and your moodiness is extreme. But, as the rhythm moves up towards the next critical day, life begins to take on a better perspective. At this switchover day you may exhibit humour that could be questionable. Over-confidence could play a part, too: you may take on responsibilities you do not really want.

Generally speaking, while the normal changeover period for the physical cycle is around twenty-four hours, at least forty-eight hours should be allowed for the emotional cycle. This would allow a clear period of time to get any possible lapses of behaviour out of the way. Driving should be avoided if it is at all possible because, when pessimism prevails, the danger of accidents looms large on the horizon. Putting yourself into stressful situations does not help either.

Of course, most of us are committed to certain courses of action or behaviour by virtue of our occupation or circumstances and we cannot be expected to shut up shop for such a long time on a regular basis. Yet in some countries this is more or less exactly what does happen. Drivers are asked to work indoors for the duration of critical days or, if allowed to drive, are asked to display a little flag that shows the rest of the world their biorhythmic state. In some Japanese cities, where biorhythms are almost a way of life, the taxi-drivers show almost fanatical enthusiasm for the subject and it is now a common sight for these little flags to be seen flying from various vehicles.

The Question of Free Will

So, the cycle continues, moving into the positive stage once again and, with the accent on sensitivity, it is obviously wise to take the trouble to chart your emotional rhythm. Even if nothing happens, and it may not for years, it is a reliable guide for avoiding the possibilities of emotional conflict. However, it must not be accepted that because you are to experience a critical day you must do this, or that you must do that. Arguments are frequently employed to prove that scientific behavioural predictions are impossible: how can we attempt to explain or predict something that, first and foremost, is subject to free will?

Man chooses his own way of life, usually to conform with his surroundings, but not always. Non-conformity at least accounts for the con-men, business geniuses, creative artists, heroes and heroines. We all have the free will to choose our actions, and the vast majority of us conform by finding the most comfortable niche in our society and staying there, only rarely straying from it when we feel the urge to do so, or by invitation.

However, in the defence of this niche we are at our most vulnerable or, at the very least, exhibit a pattern of behaviour that observes what is now established as normal, below par or above par. Couple this social conformity with the results of research into behavioural patterns and we all fit somewhere. Some will stay out in the cold a little longer than others but, eventually, they too, will return to the fold to be counted.

In competitive events of all kinds the most natural thing is to want to win, and the nature of the winning reflects the behaviour of the individual. There is the outright competitor who has to win and employs all the skills at hand to do so. At the other end of the line we have the rank outsider who hasn't a chance unless ... The emotional rhythm in the plus phase will help both contenders. Both have equal chances at the starting line, but it is performance that counts. The entertainer performs far better when his rhythm is in the positive stage. Small, personal nuances can be expressed far more effectively, particularly important to the comedian practising his difficult art.

Using the Positive Stage

Engaging in teamwork of any kind requires a certain amount of acting ability, particularly if personal preferences have to be put to one side in order to concentrate on the issues at hand. Whether involved in sporting events, crime detection, selling, advertising, marketing or simply being part of a team in an office, a certain amount of sublimation of natural desires and preferences has to be exercised for successful results to be achieved. This is best done when the emotional rhythm is in the positive stage. Even the television newsreader or programme presenter is less likely to fluff his lines at such a time and stay on an even keel.

If you are thinking of proposing to your current love, you might be forgiven for first establishing your mutual biorhythm status for the best results: you could be in for a big surprise. It is frequently found that married partners are often at total variance in their respective rhythms, but this will be dealt with more fully in the chapter on compatibilities. If you are the one doing the asking, make sure that you are in the positive phase at least. You will be better able to cope with the results of your proposal then.

When planning large-scale business matters, do so in the positive phase of the emotional rhythm. This will enable you to convey your enthusiasm to others, to be on the ball, and be able to cope well in question and answer sessions. Even if you are physically down, biorhythmically speaking, you can still take part in a wide range of sports and enjoy the experience if in an emotionally plus phase. All you have to do is remember your physical ability for the day in question and do not overdo it.

Allowing for the Negative Phase

When emotionally under par, in the negative stage, there is still no reason why you should not take part in any kind of event, provided you allow for your limitations as indicated by your personal chart. During the approximate 48-hour period of a critical day, taking part in sports which include an element of danger or which need split-second timing are not recommended, however. Stock-car racing, motor sports, motorcycle events, horse trials and similar events should be avoided if at all possible as you are more likely to be prone to error: but if you want to kill yourself that is entirely your affair. The only trouble is, with competitive sports, you may take someone else with you, or be responsible for the demise of others because of your inability to cope with situations as they arise.

Biorhythmic Pre-knowledge of Others

Biorhythms will not predict the outcome of any event, of course, but they will indicate the potential you have at your disposal for any given day. How you use this information to dictate your subsequent actions will be entirely up to you to decide. The advantage of biorhythmic pre-knowledge is that it not only indicates your particular disposition for the day, but that it must do so for everyone else also.

This fact can prove an unprecedented gift to the astute in

the sales and promotion business. I know of one sales manager who had biorhythm charts prepared for all his salesmen and his biggest buyers. A little judicial juggling here, some careful timing there, and the graph on the sales chart went soaring sky-high. It must be to your advantage to be aware of the other person's weakness, and can then capitalize on it. This is not an unfair advantage; it is utilizing biorhythms intelligently to get a little more out of life, and it is open to everyone else to do so.

If you suffer from high blood-pressure, you will want to avoid deliberately taking risks on an emotionally critical day. If sensible, you are likely to keep a curb on your tongue if a slip can easily affect business relationships in important policy matters. You should concentrate on trivia and tidy up loose ends at those times when you know that substantial concentration on more important matters is not so easy.

Good Timing and Good Sense

If you lack perception at the best of times, you could improve your performance by more than 50 per cent if you learned to time your affairs so that they nearly always favour you. One of the worst possible places to be is at the office when the annual Christmas party takes place. It not infrequently portrays people at their worst. If you are easily embarrassed and it is not going to be a good day for socializing, you have two choices. You can make a reasonable excuse and not go, although this may be wrongly interpreted as a slight by colleagues and boss alike; or, you can attend but keep as low a profile as possible, slipping away at the earliest opportunity.

The sudden freedoms involved in such situations, before alcohol adds to the confusion, will often expose people exhibiting almost perfect natural biorhythmic responses. A simple check on individual biorhythms would be helpful in such situations to ensure that all get the best from the event. Naturally, not everyone's biorhythms will be ideally phased at the same time, but a careful and considerate host will ensure that no one gets left out of the fun entirely. We often find that we can get on with certain people even when our biorhythms indicate that we should not be in company at all. Emotional compatibility with others is dealt with more fully in the chapter on compatibility, but it is worth noting at this point times when more caution than usual should be employed.

Sensitivity to atmosphere and people needs careful handling: often a careless word or deed at the wrong moment can prove not only very embarrassing, but may stay in the memory for longer than you think. In the negative stage of the emotional rhythm one is particularly prone to such awkward situations. The simplest advice is to remember what could happen, and behave accordingly. We are then well on the way to automatically correct behaviour in our relationships with others. Good behavioural patterns will be noticed by those who matter. People who have learned such a lesson may be singled out for promotion, receive recognition earlier than could be reasonably expected or may be required to perform a specific task that they may not ordinarily have thought they could handle.

I have often been struck by the attitudes of interviewers in employment agencies. Sometimes they appear well-suited to the job, at others they seem totally out of place. The best times to be on duty could easily be assessed from a simple biorhythmic check, other tasks could be quite easily handled on a temporary basis on inauspicious days.

Similarly, a teacher whose biorhythms are not conducive to giving direct instruction could always utilize such periods for setting written work, thus averting possible loss of face when the class plays up. School discipline eventually improves as a result; for there can be little worse than a teacher who is forced into a display of temper or a slanging match with a pupil.

These are just a few of the lessons to be learned from simply keeping a watchful eye on the emotional cycle alone. Utilizing the three rhythms together improves performance all round, as I shall demonstrate in later chapters. Those who perform well will be noticed: promotion invariably follows for those who demonstrate ability of any kind, especially when it involves human relationships. And none of us can live without others.

4

THE INTELLECTUAL CYCLE

The third of our cycles controls all intellectual responses: the powers of reason, perception, judgement, acuity and plain common sense. The positive phase, which begins with a critical day, accentuates these powers for the good – you shine mentally. Observation is at its peak, simple mental problems are dealt with easily, often without realizing it.

Mentally, you will feel in fine fettle and wish to exercise what you may consider to be an under-employed mind. There may be a tendency to make lighthearted calculations from your observations or test your memory with mental games for the sheer pleasure of flexing the mind.

In fact, a book of puzzles at this stage of your cycle is highly recommended because if the mind is under-exercised it quickly becomes bored. A highly active intellect left to stagnate for any reason can create social problems. Unintentionally hurtful, cutting, sardonic remarks or justifiable criticism which are really best left unsaid can quickly spoil an otherwise good relationship.

The Positive Intellectual Phase
It is doubtful whether we ever use our brains to their full

capacity, and this long rhythm of thirty-three days, if not correctly understood and properly utilized, will almost certainly result in the under-employment of a valuable asset. The mind is a funny thing: the more it is used, the better it flourishes. It is like a car-engine, the more it is used correctly, the more it will serve you efficiently. Continual short journeys will choke up an engine eventually; a similar result is achieved by the mind under the same circumstances. In the positive phase of this cycle, then, the best thing is to keep the mind as fully stimulated as possible.

Students will find that they will probably learn more in half-an-hour of study during this phase than they do in a whole day during the negative stage. Creative pursuits proceed more easily: ideas flow and perception is at its highest point. The senses are more acute and responses are at a higher level. Debates and discussions go well, conversation is easy and flourishes. Because the mind is in such good shape, during the space of an hour or so the amount of subjects covered over the after-dinner brandy, coffee and cigars will surprise you.

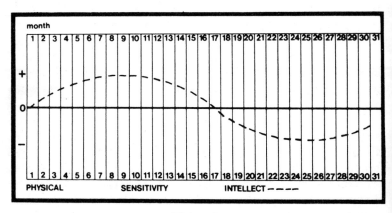

Figure 3

At this stage of the intellectual cycle, ambitions can receive a boost, especially if you are trying to impress people who matter. You shine at meetings and interviews, and can display your full potential successfully. Your powers of concentration will be at their best, so it is a good time to take tests and exams. Planning events of any kind will seem easier; small details which may ordinarily escape your notice under different circumstances are remembered and dealt with.

Money problems beset us all, but if you do have to sit down to sort things out, now is the time to do so. Tackling your budget will seem easier – you may even find that more can be put into the savings account than you first thought possible. This is also the best time to start a new job. All those fresh faces and new routines that will have to be mastered will be handled with an expertise you may not have thought you possessed. Any potentially difficult social problems are considerably eased at this stage of the cycle because the mental processes are working so well that there is sufficient time to remember and deal with details.

Intellectual Critical Days

On a critical day, however, intelligence becomes muddled. You may wake feeling a little fuzzy in the head and, as the day progresses, find that it becomes difficult to express yourself properly. An intellectual critical day is not the one on which to make important decisions. Judgement and common sense will be impaired, and memory can play tricks. Travel could prove a problem, especially if on unfamiliar ground, as this would tend to emphasize any possible confusion which you were experiencing.

In the intellectual rhythm, critical days tend to have a slightly longer effect than in the physical cycle. As with the emotional rhythm, it is wisest to allow at least forty-eight hours before you can consider yourself really free of the effects. So, remember, if you have any important meetings scheduled or need to make skilled decisions on or near a critical day, it would be far better to postpone the affair if at all possible. If the matter cannot be delayed, at least try to remember that you are more likely to be error-prone on such occasions.

You may, of course, feel that nothing is wrong at all. You

could sail through an examination and finish early; but you may also find that you forgot or overlooked important points when you go back over your work to check for errors or omissions. So, take your time; you might not want to appear 'slow' in certain types of company but, if you have made an error, it is far better to discover the fact before any real damage is done.

If you are employed as a pay clerk, for example, and overpay someone, the chances of getting the money back are remote. If you are held responsible for irregularities in your cash dealings, it could cost you more than just cash. Your reputation, ability, honesty and integrity may all come under scrutiny, yet all you did was to make a simple error that anyone could have made in similar circumstances.

The Negative Phase
In the negative stage of this rhythm, the mental processes are slowed down: perception is dulled, concentration is lacking and even the simplest mental tasks seem to require an enormous effort. Sometimes these ups and downs in your mental capabilities remain unnoticed from one year to another because they may be only very slight. The college lecturer, for example, who is continually flexing his mental muscles may only exhibit a slight difference in ability. A fluffed line here, or a short pause there while his memory suddenly seems faulty may be all that is apparent. On a critical day, he may inadvertently address someone by the wrong name or title, or perhaps repeat himself once or twice.

A sportsman, however, who lives a more physical life and is not called upon to use his mind in the same manner as a professional teacher, may notice that his concentration has failed completely. Curiously, the negative phase of the intellectual cycle does not seem to have a very great effect on driving. It appears that most people drive cars 'automatically', and that emotional reaction is far more deeply involved with this ability than intellectual acuity. Obviously, responses will be dulled a little, but once one has actually learned to drive, one is inclined to physically and emotionally 'absorb' driving techniques. Thus, mental processes are not used as much as one would naturally suppose. Motorcyclists, however, have to continually use their mental judgement while driving and would be well-advised to keep a weather-eye on their intellectual rhythms.

It is a generally considered opinion of biorhythm exponents that the second half of intellectual rhythms, the negative phase, should be utilized for reviewing work wherever possible. Whereas, new tasks are better tackled during this cycle's positive stage: actors and actresses, for example, would find it better to go through their new scripts during this period. The mind is less perceptive and does not want to become too involved in new material during the negative phase and is happier just ticking over. So, it is an ideal time for reappraisal, for going over information or material already gathered.

Crime detection often peaks in this phase. The painstaking care with which a detective has to piece together all his evidence is well-known. He may get his inspiration during the positive phase, but will find it better to mull over what he already has when in the negative. However, should he have concluded his investigations, it would be far better to postpone arresting his prime suspect on an intellectual critical day. If this is not possible, it would be wise to make absolutely sure of his facts before moving in.

We hear of so many cases being lost because of a legal technicality in the criminal's favour that it would be interesting to know if any research has been done on this point. At present, the only information to hand is that crime occurs 4.7 per cent times more frequently on intellectually negative days than on positive ones.

Serious crimes, such as murder, rape, kidnapping and armed robbery, take place most often when the subject is in the positive stage of the physical cycle and the other two cycles are in negative phase: this is the stage when an individual is most capable of great violence. Obviously, we can only view crime problems in retrospect, but a glance through the records does tend to support this theory, although there is insufficient data available to confirm it beyond all reasonable doubt.

Politicians, in particular, have been found to be susceptible to fluctuations in the intellectual rhythm. Constantly in the public eye, they tend to be acutely aware that they are public figures and, for the most part, behave accordingly. But, if they are going to make mistakes, it is much more likely to be on the critical days of the intellectual rhythm than at any other time. Politics is a mentally-orientated way of life. It is also a taxing business and many a politician has, at inopportune moments, well and truly put his, or her, foot in it when the intellectual cycle has been in an unfavourable phase.

Awareness of Potential

The mini-critical days, at day nine in the positive and day twenty-six the negative, tend to display the subject's extremes of mental prowess. When in the plus stage, mistakes or overconfidence frequently occur – almost as if the person concerned has neglected to do his homework. But brilliant streaks may be shown during this phase, too. The opposite is true in the negative phase: it almost seems as if the mind cannot cope with any situation. These are extreme examples, of course, but they do serve to illustrate the way in which biorhythms can be employed as guidelines for living.

This does not imply that one cannot live without consulting a biogram, nor that accidents will always occur on critical days; accidents can and do occur at other times. But it is wise to be fully aware of your potential before embarking on any dangerous or potentially dangerous activity and to try and restrict such undertakings to days when your rhythms are in favourable conjunction with each other. For instance, you are most likely to survive trouble if your physical and intellectual cycles are positive and your emotional rhythm is negative than at any other time.

The intellectual cycle is the least studied of the three biorhythms, partly because most research into the physical manifestations of biorhythms seems to be connected with the physical and emotional cycles. Nevertheless, the intellectual rhythm should be given the same importance as the other two because life is a series of logical steps.

Biorhythms – A Key to Success

Nearly everything you do, if you really stop and think seriously about it, is governed by social obligations which, in turn, are based on survival. If you look around, everyone is doing the things you do: obeying the same impulses, observing the same rigorous codes of behaviour. Admittedly, each of us interprets our way of life in our own individual style and some people stand out more than others – though not necessarily better than others. There are those who have given up altogether: the tramps, the down-and-outs, the unfortunates who have suffered some blow to their once normal lives and ceased the struggle.

Those who really stand out are those who seize each opportunity as it arises. Who is to say that they have not used their brains to achieve their positions which you may envy? Many prefer to opt for the argument that these folk got where they are because their faces fitted or because they had contacts in the right place at the right time and that ability had little to do with their success. This could be true to a degree, but sheer ability got them noticed in the first place. The ability to think, reason and make decisions based on a predetermined course or by following their natural desires.

The naturally ambitious will nearly always succeed in using their mental powers to gain things they wish to achieve. The intelligent use of biorhythms could have you up there alongside them, vying for the same things on equal terms. Naturally, biorhythms cannot give you everything on a plate; that simply is not possible. But abilities which you do possess can be strengthened by timing your opportunities to the best possible advantage.

You want to succeed, most of us do. Your biorhythms, used intelligently, will help you along the all-important road to success.

5

COMPATIBILITY

We all want to get along as best we can with those around us, but this can sometimes be uncertain. We may find ourselves behaving slightly irrationally towards someone we have just met for the first time, something does not seem to quite 'click' into place. In effect, our relationship does not get off to a very good start. As time goes by, we find that there is something about the association that prevents it from really getting off the ground, no matter how hard we try. At other times, an immediate rapport is achieved. We find that we think alike, act alike and can generally have a stimulating time in each other's company.

By comparing our biograms we may find the solution to these apparent discrepancies in relationships. After all, it is a logical step to propose that what biorhythms do for us, they must do for others. Reactions to personal biorhythms vary from individual to individual, of course. Yet they can provide guidelines to how we can most successfully relate to others, and the system used for comparisons is easy to follow.

If you look at Figures 4 and 5, you will immediately observe the similarity between them. They both have intellectual

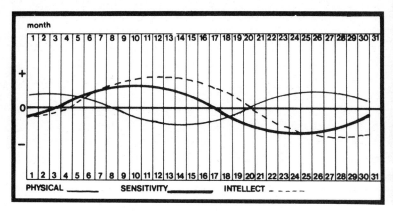

Figure 4 A

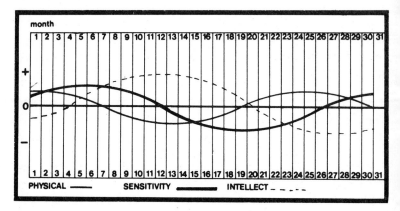

Figure 5 B

critical days occurring on the 4th and 20th of the month. The first subject, whom we will call A, has physical critical days on the 8th and 20th, while B has his on the 7th and 19th. Their emotional critical days differ a little more. Subject A has his on the 3rd and 17th while B's occur on the 12th and 26th. Their intellectual rhythms, therefore, are exactly in phase; they are one day apart in their physical rhythms but are nine days apart in their emotional cycles.

In such a case, therefore, it is reasonable to expect these two people to think alike and act alike, but that there may be a slight difference in their emotional approach to life. They are, in fact, very good friends. They hit it off the day they met and their association has gone from strength to strength ever since.

The subjects of Figures 6 and 7 are quite different. We will call them C and D. They both have emotional critical days on the same day, but their cycles are in opposing stages. Subject C's positive phase runs from the 6th to the 20th, but D goes into the negative stage on the 6th and switches to the positive phase on the 20th. C has her critical physical day going into the negative phase on the 10th and her positive critical day on the 22nd. D's cycle is the reverse of this: her rhythm goes up on the 10th and swings down into the negative phase on the 21st. Intellectually, their biorhythms are virtually one day apart. C experiences her first critical day on the 8th and switches to the negative on the 24th, D moves into the negative on the 7th and changes back into the positive phase by the early hours of the 24th.

Compatibility Ratings
In biorhythmic terms, A and B have a total compatibility assessment rating of 85 per cent, which is unusually high. This is made up of 91 per cent physical compatibility, 64 per cent emotional and 100 per cent intellectual. The total compatibility ratings for C and D, on the other hand, is only 7 per cent. This unusually low figure is made up of 4 per cent physical compatibility, nil per cent emotionally and 3 per cent intellectually. These charts are taken from life and are of four people who have to get on with each other. Although A and B are just good friends, C and D have to work with each other.

Before C and D were aware of their compatibility rating,

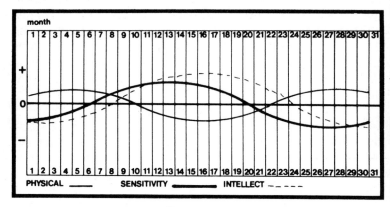

Figure 6 C

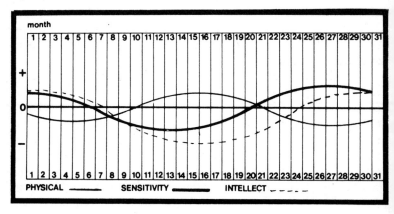

Figure 7 D

there was a definite feeling of antipathy between them. Now that they are both aware of a possible reason for this feeling, they tend to get on far better than they did originally, although their basic personality differences are still apparent. They have learned to get along with each other by making simple adjustments to their own personality traits while, at the same time, recognizing the problems that may arise between them.

An easy method of checking the compatibility rating is by calculating the difference between the individual cycles in days, each day representing a proportional difference in the cycles. These calculations are made for each rhythm individually, then added together to give an overall figure. If this total is divided by three, the number of cycles involved, the resultant figure represents the overall compatibility rating. The important point to remember is that this compatibility factor remains constant, no matter what the individual reading for any particular day may be.

In Figures 8 and 9, on the first of the month E is on the fourth day of the physical rhythm in the positive stage and F is on the eighth day, the difference being four days. If we look at the chart of compatibility percentages in the physical rhythm, we will see that four days apart represents 65 per cent. On the second day E's rhythm will be on day five and F's on day nine: still four days apart.

In the emotional rhythm, E is on day one on the first of the month and F is on day eight, the difference being seven days, or 50 per cent. The following day they will have increased their stage by one more day, but the compatibility factor remains at 50 per cent. In the intellectual rhythm, E is on day two and F is on day fifteen on the first of the month, representing only a 9 per cent compatibility factor.

If we add the three factors together, 65 per cent for the physical, 50 per cent for the emotional and 9 per cent for the intellectual, we arrive at a gross figure of 124. We then divide this total by three (the number of rhythms) to arrive at an overall assessment figure of 41 per cent. This overall figure also remains constant.

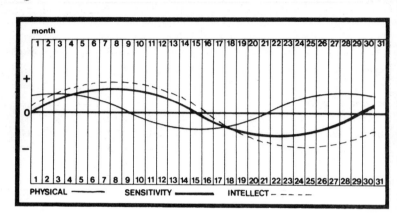

Figure 8 E

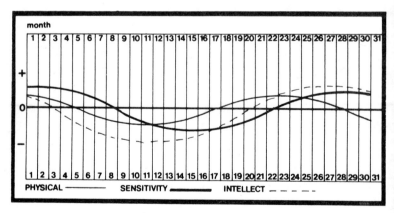

Figure 9 F

Days Apart in Cycle	Physical Cycle (per cent)	Emotional Cycle (per cent)	Intellectual Cycle (per cent)
0	100	100	100
1	91	93	94
2	83	86	88
3	74	79	82
4	65	71	76
5	57	64	70
6	48	57	64
7	39	50	58
8	30	43	52
9	22	36	46
10	13	29	39
11	4	21	33
12	4	14	27
13	13	7	21
14	22	0	15
15	30	7	9
16	39	14	3
17	48	21	3
18	57	29	9
19	65	36	15
20	74	43	21
21	83	50	27
22	91	57	33
23	100	64	39
24		71	46
25		79	52
26		86	58
27		93	64
28		100	70
29			76
30			82
31			88
32			94
33			100

Figure 10 Compatibility Table. All figures are quoted to the nearest whole number.

Key Factors

It is at this point that there is a difference of opinion between biorhythm experts. Some feel that this overall figure should be used as a guide only, while others think that it is a reliable pointer to comparing personality. In either event, it has been noted that whichever cycle shows the highest percentage in the total will feature most strongly in the relationship.

Thus, between A and B the figures were 91 per cent, 64 per cent and 100 per cent. The highest figure, therefore, is in the intellectual cycle. The relationship between these two people is accentuated by their almost uncanny ability to read each other's minds when working things out. They have the same basic approach to life and display exactly similar instinctive perceptions. Emotionally, they tend to share similar feelings on a wide variety of subjects and believe in the same emotional ideals.

In the case of C and D, it is a completely different story. The totals were 4 per cent for the physical cycle, nil percent in the emotional and 3 per cent in the intellectual rhythm. Not a lot to work with here when trying to assess the overall key factor in the relationship because the figures are so small and so close. In fact, there is no key factor, these two people simply have nothing in common.

Partnerships

The higher the compatibility factor in the physical rhythms, the more readily you will undertake those tasks requiring physical effort in unison. Sports and recreational activities need to have both or all parties acting more or less harmoniously for the duration in order for them to be enjoyed to the full. Put two people in a canoe and send them off down the river, or put two people at the base of a mountain and let them climb it. In either case, it is not going to be particularly pleasant for the one if the other cannot or will not pull his weight.

The more out of step the biorhythms, the greater the compensatory approach that will be needed because at least some sort of harmony must be struck if there is going to be a good day. Compromises made at an early stage will go a long way to achieving a worthwhile all round day.

One of the partners may be in the positive phase of the physical cycle when the other is nearing the end, on a critical day, in the negative phase or just beginning the physical cycle. In such a case, the former partner must be allowed to take the lead, with the other bringing up the rear as best he can in the prevailing circumstances. Provided that each allows for the difference in phase and understands how this will affect their individual performance, the day should go well.

Group Compatibility

But compatibility does not stop at just two people. The rating between three, four or any number of people can be worked out, given time, and used to achieve more successful relationships. This assessment will allow the parties concerned to maximize their available potential. Their thinking, therefore, must also become more positive, with improved results all round.

For example, if we look at the problems confronting a football team manager when selecting his players for the Saturday match, the intelligent use of biorhythms would probably go a long way in helping him utilize the best possible players at their time of maximum potential. The first reaction is to choose eleven players who are at the peak of performance as far as their biorhythms are concerned, but this would not necessarily achieve the best results.

A reasonable amount of non-compatibility must exist between any group of people, for a variety of reasons. Most clubs have far more players than is generally realized; they need to allow for substitutes, reserve teams and trainees, and will have certain standards to maintain. So, the first step is to see how each individual gets along with the others, on and off the pitch. The compatibility between the players and the manager should also be checked.

Having established this part of the exercise, the next step is to select from the club's star players as far as possible. In the unlikely event that all of them get along with each other perfectly, the selection of the best possible team, biorhythmically, would then be easy of course. Yet it would still be inadvisable to select all those players who were at the peak of biorhythmic positive stages for there would then be less likelihood of achieving a 'team' performance: each player would feel that he was fully capable of winning by his efforts alone. Therefore, the best selection would be on the basis of a high positive physical stage for each man with, as far as possible, the best compatibility rating between them; not forgetting to allow for substitutes.

Unless there was a marked emotional incompatibility between them, the teams's emotional phasing need not necessarily be high or even in the positive stage because there would be a greater inclination towards sharing the game at other stages. Intellectually, all that would be needed is concentration, irrespective of the cycle's phasing. So, if all the players were at absolute peak biorhythmic performance in each of the cycles, they would most likely lose the match because of their inability to perform successfully as a team.

Cricket, however, calls for a different set of skills and would need a different approach towards the selection of a successful side. The batting team has to have a steady eye and so, intellectually, needs to be in the plus phase. Also, they require the physical ability to make strokes that score, so their physical rhythms would need to be auspicious. A bowler needs to be perfectly attuned, so his emotional rhythm would need to be good and, of course, in order to be able to consistently deliver balls powerfully and accurately, his physical rhythm would need to be considered also.

Tennis requires stamina, vigour and concentration, so all three rhythms should be in positive phase to be successful. But the emphasis is on accuracy, therefore the emotional and physical rhythms are additionally important. With all professional sports, once the techniques have been mastered, the actual play becomes almost second nature, therefore emphasis is placed on the emotional and physical rhythms being in positive phase for successful participation.

In a department store, numerous people have to get along with each other and the customers under fairly restrictive conditions. Obviously, it would not be practicable to verify the biorhythmic compatibilities of such a large number of people therefore; nevertheless, the staff could benefit from knowing their personal rhythms and individual compatibility potential. After all, selling techniques aside, individual sales staff who experience friction in certain relationships must find their selling ability curtailed as a result.

Working Relationships
Often the cause of a poor working relationship stems from unsuitable management. Irrespective of the natural banding together of junior staff against any oppressive superior, the potential working atmosphere may be easily assessed by a simple comparison of all their charts.

A						
15	B					
64	91	C				
47	85	79	D			
70	86	89	92	E		
30	100	94	86	80	F	
10	18	24	12	15	33	G

A = Department Manager
G = Odd Man Out

Figure 11

In Figure 11, A is the department manager, and you can see that he barely gets along with two of his staff and not at all with the others. Higher management could solve this particular problem by transferring A to another department to head a different sales team with whom he was more compatible, and by placing another manager in charge of the original group. Increased sales would inevitably result. Another improvement which could be made would be to place G, the odd man out in the original group, in another department: to mutual advantage.

Translate this sort of problem into a factory environment and the inherent dangers become quickly apparent. Conveyor-belt production techniques are monotonous and can lead to boredom and irritation, frequently resulting in accidents. Also, employees may have to share machinery for short periods. Again, incompatibility and the need for bursts of absolute concentration can lead to accidents. And when accident rates rise, so do insurance costs, and the tempers of management. Returning to Figure 11, if G was the foreman in such an environment, the turnover of staff would probably be very high indeed.

Among library staff, it has been found that although almost everyone gets along in their intellectual phasing, their emotional and physical ratings can be uncommonly poor. This is a good example of working conditions attracting people who have the ability to do a specific type of job, but not necessarily the ability to get along with each other.

The incident rate of incompatibility is often inordinately high in a typing pool staffed by young girls. The supervisor must, therefore, exercise discipline combined with understanding. Physical compatibility is not of great importance here, but emotionally and intellectually there is a need to utilize wide discretionary powers. This could be achieved by allotting certain types of work only to those who show an aptitude for that specific task. Of course, this may create other problems, but as long as the staff are given an adequate explanation a biorhythmic compatibility assessment of the employees involved will improve output to a significant degree.

Improving Relationships
On the whole, most people will try to improve their relationships with others if explanations and guidelines are offered. You may not really understand why you have never

got along with him yet always get along with her, but you are probably willing to try and improve a poor relationship even if there is no guarantee of success. Biorhythms may reveal a potential solution to the problem, although it must be remembered that compatibility is subject to personal behavioural patterns and the prevailing conditions of the moment.

The percentage figures are constant and, for the best results, personal biorhythm charts should be calculated so that mutually advantageous times for joint activities can be arranged. The overall assessment figures should be regarded as a guide only – the individual figures are the more important.

Physical Percentages

When assessing the physical rating, 100 per cent is fine for all activities requiring joint participation. A figure of around 75 per cent will require the temporarily stronger partner to allow for the weaker partner's inability to be as active. 50 per cent or less indicates that the difference requires the good judgement of both partners to time events for mutual satisfaction, leading to a better understanding and more rewarding relationship.

Where several people are concerned, greater allowances need to be made. Remember, with a group of four or more, at least one member is likely to be present purely for the sake of the activity involved. For example, three fanatics and one player along for the ride might make a lot of difference to the enjoyment obtained from a round of golf.

Emotional Percentages

In the emotional cycle, a 100 per cent rating is fine for most relationships except for long or close partnerships such as marriage or inter-family associations where too similar rhythms can lead to tension. In very close relationships, having the same ups and downs and the same critical days can be boring and cause almost as many problems as completely incompatible emotional rhythms. Too little stimulation and a relationship may stagnate and, if both partners are in a bad mood on a negative critical day, heaven help the situation. From 45 to 65 per cent is about the best rating for married partners because this differential will provide a constant stimulus owing to the fact that their rhythms will be slightly out of phase.

An emotional assessment of below 40 per cent is poor, however. Both partners will need tact, good timing and understanding in any situation requiring emotional rapport; a careful study of the individual charts will reveal the best time to stage such activities.

Intellectual Percentages

A 100 per cent rating produces the best results in all relationships where the intellectual rhythm is concerned, although a lack of stimulation could be experienced because of the parties' 'think alike' ways. Around 75 per cent is a good figure because one partner will complement the other when the cycles are at variance. The higher level of perception of the one partner will offset the under par thinking and reasoning of the other. An assessment of 50 per cent or lower shows a contrasting intellectual approach, requiring tact and diplomacy that could stretch forebearance at times. But, if both partners allow for the difference in their perceptive levels and, again, use their personal charts to time the best moments for those important joint decisions, problems can be avoided.

The higher the overall compatibility rating, the better the relationship is likely to be, although it may be coloured by the rhythm with the highest percentage. For example, if the physical rating is the highest, it probably means that the relationship is based on both partners' enjoyment of each other's physical presence. If the highest rating is on the emotional level, it means that the partners are temperamentally suited; and if the intellectual rating is the highest, that they are mentally attuned.

Knowing the Good and the Bad

When starting a business relationship with a friend, it would be wise to check your compatibility rating. If intellectually high, but low on the other two, do not let it worry you too much; work at the weak points in the relationship and your business projects should prosper. Even if the assessment figures are poor all round it does not mean that you cannot succeed. But you will have to work a little bit harder at maintaining the status quo in those areas where you are at widest variance.

Young couples considering marriage may well find themselves dismayed by a poor emotional assessment but, as has already been indicated, this has proved to be an almost standard result in long-term marriages and other successful close partnerships.

The teacher/pupil relationship, driver/conductor, doctor/nurse, nurse and patient, too, if it is a long-term affair where the patient may be looked after by a specialist nurse, all may have their compatibility assessed for a better mutual understanding. As improvements start to show, potential physical, emotional or intellectual problems can be alleviated by long-term planning based on biorhythmic knowledge.

Insight into your good days and bad must help all your relationships.

6

OTHER CYCLES

A few years ago, in 1972, a popular singing group, the New Seekers, came out with a record called simply *Circles*. The record has since become a favourite of mine, but it was really the lyrics which immediately struck me as apt, for they stated that there is an all pervading rhythm to life. How true this is.

Cycles feature dominantly, not only in the biological study of man and his behavioural patterns or even in the history of the race, but also in practically everything else that we care to consider. There are ultradian cycles, rhythms of less than twenty-four hours duration; circadian rhythms, those of one day's length; and infradian rhythms which last a lot longer, as much as years in some cases.

The study of cycles can be quite fascinating and deserves a mention here because there exist some astonishing rhythms. The average man has probably never considered most of these cycles, yet they affect him in some way almost every day of his life, whether asleep or awake. In each case, I am referring to documented studies: either proven under laboratory conditions or, where possible, observed in the natural state.

This book is about man and his behavioural patterns, so we will start with him.

'Night' and 'Morning' People

Apart from sleep patterns which, under normal circumstances, take place once a day, usually at night when all naturally diurnal creatures rest, man shows a definite rhythmic pattern in every aspect of his life.

The human body temperature reaches a peak during waking hours and is at its lowest point when you are asleep. As your temperature rises, so does your efficiency; conversely, you become less effective as your temperature drops. However, none of us is exactly alike; we respond to our body rhythms as individuals and not collectively. This is why some of us are 'night' people and others are 'morning' types.

One way of checking which of these two categories applies to you would be to take your temperature on the hour every hour until you go to bed: a task which would have to be spread over a period of several weeks. If you kept a graph of the readings, it would be quite easy to judge your most effective periods so that you could make simple adjustments to your way of life. You could, for example, time your maximum efforts with those periods of highest readings. The long-term effect would be to increase your personal efficiency that much more.

Everyone has a natural eating rhythm too, although most people tend to ignore this because of their social or business obligations. It would be better for some folk to take sustenance of some kind every ninety minutes, for instance; others would benefit most from three main meals a day. The constituent parts of any meal affect our efficiency in different ways. Some of us eat hearty breakfasts with a couple of smaller meals later in the day whereas, in fact, we might benefit from a light snack in the early hours with a heavier meal at about 11 a.m. The composition and times of meals need to be geared to individual requirements so, if you feel you might benefit from experimentation, go ahead.

Your heart beats 103,680 times per day and, of course, this organ is part of another cycle: the vascular system. The heart pumps blood around your body in a never-ceasing cycle that does have detectable variations in much the same way as do sleep cycles, feeding habits or other bodily functions.

And, for as long as your heart continues to beat, you will, on an average day: breathe 24,000 times; consume an average 3 lbs of food and drink about 3 pints of liquid; and will probably walk approximately two miles using 750 muscles while you do so.

Cellular Replacement Cycle

All this and other events will require the utilization of approximately 7,000,000 brain-cells per day. You may speak 4800 words in the course of a normal day. Your nails are growing too. Fingernails will grow about .000046 of an inch while your toenails will put on about .000031 inches. Hair length may increase by about .017 inches. Mental performance is normally at its peak between 1400 hours and 1600 hours, although anywhere between 1100 hours and 2000 hours can be the expected high according to the prevailing conditions, personality and emotional state of the individual.

Not only do we feel different people in the morning than we do at night, but we really are different in the strict physical sense by virtue of the continuing cellular replacement cycle.

The scientists have even measured our memory cycles. They have discovered that, besides other subtle and often unnoticeable physiological cycles, there are definite phases of forgetfulness and memory. These apparently vary from individual to individual and seem to be associated with emotional stimulation. Nevertheless, they follow a predictable pattern.

We learn that Mr Average is about 5 ft 9 in. tall and may live to be sixty-eight, while the average woman is approximately 5 ft 3 in. and should live to be about seventy. Statistics to fit Mr Average are, of course, just that – averages – yet they can and do reveal some fascinating facts.

Life Statistics

If married, you will probably make love about 3000 times and spend a total time equal to fourteen years working. Twenty to twenty-four years spent in bed is the norm. Travelling takes care of five years; dressing, washing, shaving or applying make-up account for another four years: and you will probably spend as much as seventy days just looking into the mirror! If you are a smoker you may well consume almost a quarter ton of tobacco and the average intake of food during the course of a normal life span is quite staggering when taken as a whole.

During a total period of six years just spent eating, the average man will consume: 6000 loaves of bread, 10,000 eggs, 4000 lb of butter and 20,000 lb of fruit and vegetables; and he will wash this all down with approximately 20,000 gallons of liquid. His sweet-tooth will cause him to dispose of 8000 lb of sugar, while his meat intake is about fifty head of cattle and some 300 chickens. Shopping for all that food (and other things as well, of course) will take about three years of waiting in queues.

Not content merely to delve into all our physical idiosyncrasies and all of our senses, the scientists have made other singular discoveries in their search for cycles and rhythms.

Occupational Cycles

Loosely using birth dates as the criteria, it has been discovered that occupations are high on the agenda of cyclic performance. It has been found that musicians, for example, have a better chance of being born during November, January and February than at any other time. This is not to imply that during the low period, October in this case, there is little chance of becoming successful in this field. If you want to be an architect you stand a better chance if you were born during December, May and June, than in September or October. Bankers seem to prefer to be born in August, with March as the low point. October produces more journalists and editors than does any other month and December seems most unfavourable for this profession.

Cyclic performance is noted and logged over such long periods of time that, after a while, it appears that even this occupation is subject to a rhythm within a rhythm. This is true to such an extent that it is virtually possible to predict certain phenomena with more than an average chance of success. Not only that, but it also appears that other cyclic events are closely linked, even when they share no other relationships that can be detected or determined at the time.

Wheels Within Wheels

Who would have suspected that a link between police states and temperature fluctuations could have anything to do with patterns of style in the world of art? Co-operation and the integration of views apparently fall into the same patterns as do the phenomena of war, crop improvements and palace intrigues! All these events have been found to operate in a world-wide 100-year cycle, itself a periodic phenomenon.

In the U.S.A. between 1920 and 1955, a clearly defined pattern emerged in the construction of residential buildings. This cycle apparently exhibits a 33-month periodicity. But what is really fascinating about such cyclic events is why they occur in the first place: what is the root cause?

At present, we know very little about our discoveries in these areas. We do know, for example, that there is a definite cycle governing the abundance of snowshoe rabbits in Canada (its length is about 9.6 years) and that the lynx, owl, marten and hawk populations have the same cycle, but we do not know what the cause of that cycle can be.

In the last 300 years or so, scientists and researchers have been uncovering rhythms and cycles galore. Some of these have been discovered by accident, a few by design. It may be that we are rediscovering a lost art or science; we simply do not know. Certain historical patterns repeat themselves, both in the same country and in other parts of the world. The events follow closely similar lines to what has gone before.

Cosmic Cycles

Beyond the bounds of earthly ties are the mysterious forces that compel the universe in its constant cyclic performance. We are able to predict these cyclic motions with astonishing accuracy, but can only hazard guesses as to their cause. We have come to accept the fact that with so much inter-dependence between these rhythms and cycles, if one should fail, unknown forces could abound.

If the Moon failed to rise, for instance, would the tide fail to turn too? What would happen if the day's length suddenly altered noticeably? As a matter of fact, the length of our day has altered, just perceptibly, in recent times. But the real effect, if any, will be more noticeable in about a hundred million years or so, give or take a few million. We may smile and think that there is nothing to worry about now, but suppose this phenomenon is part of a hitherto undiscovered cycle which does have an effect now?

We simply do not know the answers. But striving to understand more of what is going on in, around and about us, is undertaken more earnestly now. One day, someone is going to find the key: it could be tomorrow, it may be ten or a hundred years from now.

In the meantime, inexplicably, we discover mouse plagues occur in four-year cycles and that every commodity price so far studied also fluctuates in cycles. Glaciers melt in cycles, the number of babies born each day occurs in cycles, so does the amount of cheese that we consume.

Synchronicity

One fundamental fact of similar cyclic performances that has come to light is the astonishing synchronicity with which they interact. All cycles of the same duration actually peak and trough at the same time. This fact provides evidence of something that we do not understand at this stage, even if we do appreciate that it can no longer be classified as random behaviour.

Over thirty widely differing subjects fall in an eight-year cycle which has been studied over a prolonged period: from the mid-1780s to the mid-1960s. In each case, they have reached their high and low spots at the same time during this period. Cigarette production from 1880; lead production from 1821; red squirrel abundance from 1926; pig-iron prices from 1784; sugar prices from earlier than 1780; and the growth of pines from about 1867: what can these widely divergent subjects have in common apart from showing the same upward and downward trends? Again, we just do not know. This pattern is also repeated in cycles that are not of eight years' duration: some are 5.91 years long, and some 9.6 years and others are 11.2 years in duration.

There are mysterious forces at work. This comparatively new study of cycles and rhythms may one day allow us to predict far more accurately than ever before our own potential destiny. It opens up vast new areas of thought with which we may merely play at present. Not that we necessarily understand the new toy we have found. We are not even sure that we have only one toy, or that we have the whole toy. It may be only part of it: like a single piece of track from a model railway set.

This is an age of discoveries, far more so than the earlier part of this century. And the speed with which we are moving suggests possible answers are just around the corner, or the next one, or the next ....

7

THE RHYTHMS OF LIFE

In early November 1960, Clark Gable suffered a heart-attack. A few days later, in a public broadcast and during the course of a discussion on biorhythms, one expert stated that he thought that Gable should be carefully watched on 16 November because, if there were going to be further developments, that would be the day on which they would be most likely to occur.

Clark Gable was born on 1 February 1901 and suffered his first heart-attack on 5 November. He had his second, and fatal attack on 16 November.

If we look at his biochart for 5 November, we see that Gable was on a critical day physically and had been on an emotional critical the day before. The 16th, another physical critical day, this time switching from positive to negative phase, was also one day away from an emotional critical. On this day, Gable's intellectual rhythm was also in negative phase.

Obviously, his biorhythms did not cause Gable to die, but their condition was not very helpful either. This and subsequent examples will help to illustrate the apparent relationship between biorhythmic states and heart-attacks, or

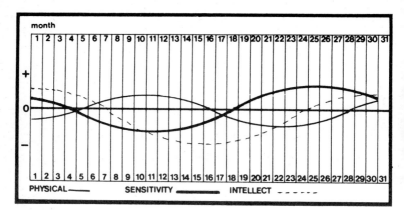

Figure 12: Clark Gable

health generally. They will also demonstrate whether certain decisions made at specific times were brilliant or catastrophic, or whether particular acts committed during poor biorhythmic phasing could be classed as coincidental or not.

It must be stressed, however, that biorhythms do not have a cause and effect in themselves, but they may be regarded as warnings or used as guides to potential situations that can arise.

Critical Days and Health

It is interesting to note that critical days, especially in the physical cycle, often coincide with poor health and even death. It has been statistically recorded that these switchover days are much more likely to be accident-prone than are the non-critical ones: so the watchword should be caution. Critical days do not, of course, cause death, but they do often coincide with it.

A good analogy is a lightbulb: if it has a weak filament, then the most likely time for it to blow is at the precise moment that it is switched on. The act of switching it on causes a burst of power to surge through the filament. Conversely, the bulb is almost as likely to blow just as it is turned off: for one moment it pulsates with life, the next there is nothing, it is temporarily weakened. Rapidly switching a lightbulb on and off in error often produces this effect: and so it appears with biorhythms.

Janis Joplin

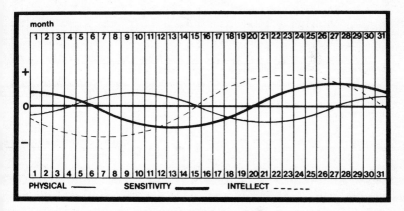

Figure 13: Janis Joplin

Janis Joplin, a popular, rock music star, died of a drug overdose on 4 October 1970. Her physical rhythm was in a critical stage and her intellectual rhythm was poor. The biogram for the date of her death implies that a bad decision was implemented when the subject was potentially accident-prone. Her body was unable to take more abuse at this point, and she left this life while still young and in her prime.

President Nasser
President Gamel Abdel Nasser died suddenly of a heart-attack on Monday, 28 September 1970. He was born on 15 January

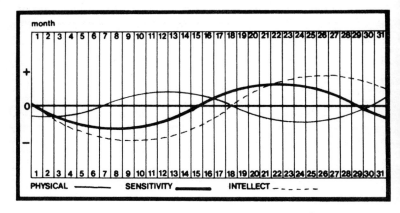

Figure 14: President Nasser

1918. At the time of his death, he was about to undergo a double critical period in the physical and emotional cycles over the next two days. If he was overdoing things at that point, his timing was certainly wrong. The use of biorhythms would have advised him that he was in a mini-critical stage in the intellectual cycle two days previously and therefore liable to believe he could cope; he was thus likely to be tempted to take on too much. With a double critical over a 48-hour period, he may not have been tempting fate exactly, but he would have been well advised to ease off.

President Sadat

The current Middle East situation may well be completely different as a result of Nasser's sudden death. President Anwar Sadat, who was born on 25 December 1918, was in a physically critical stage, emotionally under par but intellectually up when the decision was taken to go to war against Israel. It took a long time to compensate for that decision.

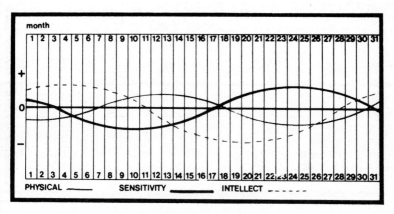

Figure 15: President Sadat

President Kennedy

But it does not always rest with politicians to make decisions that alter a nation's destiny. When Lee Harvey Oswald shot President John Kennedy on Friday, 22 November 1963, one can hardly blame the president for something that was virtually outside his control. It was not a direct act of the president himself that caused his death. However, if we look at the biorhythms for the day, there is an interesting conclusion to be reached.

It is now known, from the reams and reams of information collated in later investigations, that Kennedy had been advised of the possibility of assassination but did not pay much attention to the warnings. His biorhythms for that fateful Friday were up physically and emotionally, but at a critical stage in the intellectual cycle. Was he, perhaps, feeling overconfident of his popularity as a result? Could this have been the reason that he decided not to take the precaution of travelling in a bullet-proof car for instance?

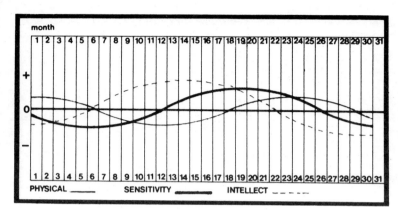

Figure 16: President Kennedy

Edward Kennedy

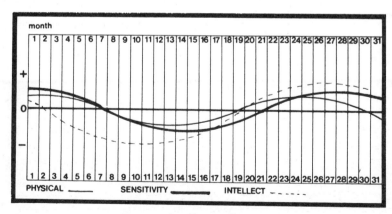

Figure 17: Edward Kennedy

Kennedy's younger brother, Edward, when caught up in the events of the night of 18/19 July 1969, now known as the 'Chappaquiddick incident', was at an intellectually and physically critical point. Edward Kennedy, whose birthdate was 22 February 1932, reacted so strangely to what is publicly known to have occurred that perhaps the only logical answer lies in his biorhythmic state on that date.

Sarah Bernhardt

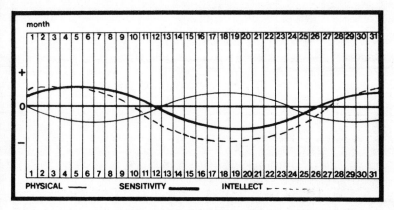

Figure 18: Sarah Bernhardt

That great actress, Sarah Bernhardt, died in a coma after a long illness which she knew to be her last. Confident and brave to the end, she continued to make as many arrangements as she could before slipping into her final sleep on 26 March 1923. Her biorhythms show that she was on a critical emotional day, had just experienced a physical critical and was about to start an intellectual critical day within a few hours.

Winston Churchill

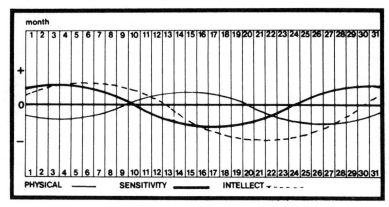

Figure 19: Sir Winston Churchill

Sir Winston Churchill died in January 1965. A glance at his biorhythms reveals that when he was first stricken with a heart-attack on 11 January, he had just had a critical physical, was four days into the negative stage in the emotional rhythm and was in the 48-hour period of an intellectual critical. On his death, he was within 24 hours of a critical in the physical cycle and the other two cycles were in negative phases.

Vivien Leigh
We find that Vivien Leigh, born 5 November 1913, died on 8 July 1967 when at a mini-critical point in the physical cycle and during the 48-hour period of an intellectual critical.

Alan Ladd
Alan Ladd, another firm favourite of film-goers, was at a double critical stage in the physical and emotional rhythms when he died on 29 January 1964. He was born on 3 September 1913.

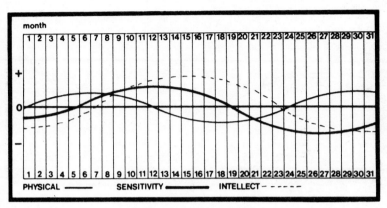

Figure 20: Vivien Leigh

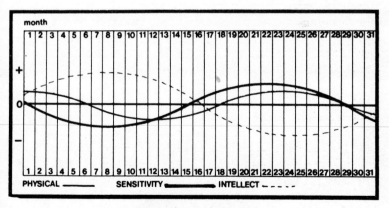

Figure 21: Alan Ladd

Michael Wilding

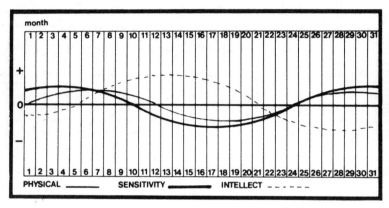

Figure 22: Michael Wilding

Michael Wilding had a short illness after a fall and died in an emotional critical period and only two days before a physical critical was due. He was born on 23 July 1912 and died on 9 July 1979.

Errol Flynn
Actor and hell-raiser extraordinary, Errol Flynn, was born on June 20th 1909 and died in 1959, on October 14th, while in a physical critical stage.

Nat 'King' Cole
A sad loss occurred in the music world when Nat 'King' Cole died, after a long illness, on 15 February 1965. He was born on 17 March 1919 and was in an emotional critical stage, with a mini-critical in both the physical and intellectual cycles, when he died.

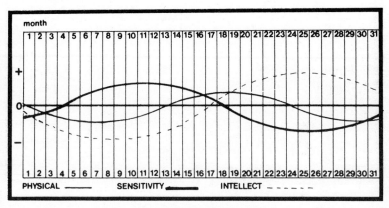

Figure 23: Errol Flynn

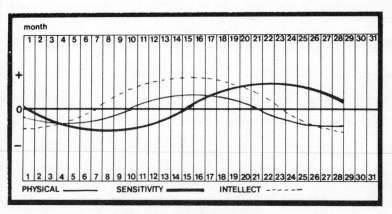

Figure 24: Nat 'King' Cole

Mario Lanza

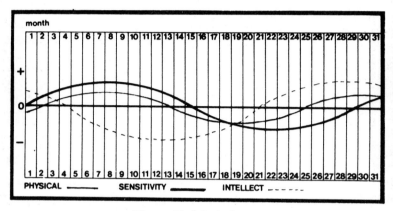

Figure 25: Mario Lanza

Mario Lanza died between an intellectual critical and two mini-critical days on 7 October 1959. He was born on 31 January 1929.

Joan Crawford

Another great star, Joan Crawford, who was born 23 March 1908, died immediately following two negative mini-critical days in the physical and emotional cycles, on 10 May 1977.

Fats Waller

Fats Waller was within forty-eight hours of a double critical in the physical and intellectual cycles with a mini-critical to add to the problem on the day he died. Fats was born on 21 May 1904 and died on 15 December 1943.

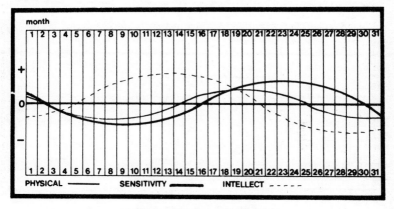

Figure 26: Joan Crawford

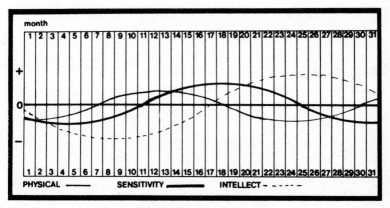

Figure 27: 'Fats' Waller

Marilyn Monroe

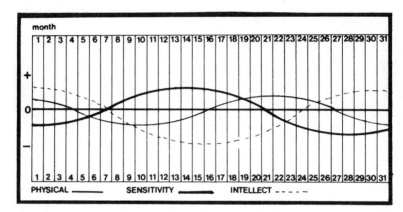

Figure 28: Marilyn Monroe

Marilyn Monroe died of a drug overdose on 5 August 1962. Her physical critical stage is apparent from the biogram, and she was only two days away from a double critical in the other two cycles. She was born on 1 June 1926.

Judy Garland
Judy Garland was another star who died of a drug overdose. Born 10 June 1922, she died on 21 June 1969 at a time when her intellectual and emotional cycles were in critical stages.

Bing Crosby
The 'Old Groaner', Bing Crosby, who was born on 2 May 1904, died as a result of a heart-attack after playing his favourite game, golf, on 14 October 1977. He was just about to experience a double critical day in the physical and intellectual cycles.

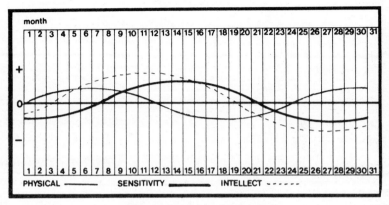

Figure 29: Judy Garland

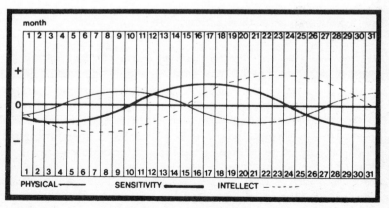

Figure 30: Bing Crosby

Elvis Presley

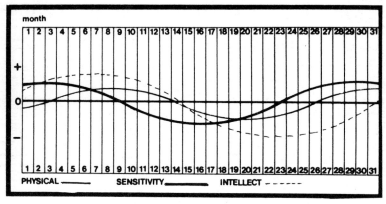

Figure 31: Elvis Presley

Perhaps the greatest entertainer of his time, Elvis Presley, died on a mini-critical day in the emotional rhythm following two critical days in the intellectual and physical cycles. He was born on 8 January 1935 and died on 16 August 1977.

These are just twenty examples of biorhythms revealing a possible explanation why these famous people behaved or died as they did. Would Mario Lanza, for example, have survived if his biorhythms had been different? Would Bing Crosby's heart-attack have killed him if his rhythms had been in better phasing? And would Marilyn Monroe have been able to overcome her deep depressive state and not have overdone things if her cycles had been differently phased?

Anticipating Dangerous Periods
Biorhythms can be utilized to anticipate the potentially dangerous periods in life, however healthy the subject may appear at the time. It does not necessarily follow that if you

have a heart-attack or are suffering from a disease that is known to kill, the next critical day will be the fatal one. Far from it. Ideally, the patient's biorhythms should be displayed at the foot of the bed alongside the other charts used in hospitals and sick rooms. At the critical biorhythmic stages, an additionally alert watch could be kept on the patient ... just in case.

In sports and athletics, biorhythms should be mandatory, especially in such sports as motor-racing, boxing and other events where physical as well as intellectual and emotional balance is of paramount importance. Benny 'The Kid' Paret was knocked out by Emile Griffith on 24 March 1962. Sadly, after the Kid lost consciousness, he never woke again. Paret was on a triple critical day, a highly dangerous situation for a sport such as boxing. He went into a coma and died on the next critical day: his body just could not face up to such an onslaught and, in its weakened state, finally succumbed. Muhammed Ali had his jaw broken during his fight with Ken Norton on 31 March 1973. Ali was in a double critical biorhythmic phase on that day. Coincidence again?

Let us not forget the referee in such instances, either. He, too, should be in the correct frame of mind to be able to do his work adequately. When certain unexpected decisions have been announced in the ring or on the playing-field, it may have been due to poor biorhythmic phasing. In some circumstances, a referee who allows a fight to continue for a few more minutes could make a world of difference for the loser if he is taking more than the usual amount of punishment.

A racing motorcyclist has little chance when he comes off his machine at speeds of over 100 miles an hour. And it may not have been his fault in the first place. A simple biorhythmic check may prevent accidents in such sports which claim so many lives each year.

The list of motor-racing fatalities is probably only too well-known. The intensity of concentration needed and the physical stamina required for participation in this sport would make the introduction of pre-race biorhythmic checks on the drivers an interesting exercise. But, although it would be all very well to discover a potential explanation for why a tragedy had taken place, as fatalities often include members of the public as well, every possible safety precaution that could be made, should be.

We frequently hear or read of incidents where a simple mistake has cost a limb or a life, whether as a result of mountaineering, pot-holing or merely crossing the road. It is not possible, obviously, to say whether the use of biorhythms would have made any difference in such situations unless the relevant information has been recorded and is available. We can trace the lives of famous people because they are in the limelight and everything that they do is news. But when your neighbour has an accident, it is usually only of local interest.

The business tycoon sitting in his office is just as prone to making poor decisions when his biorhythmic phasing is poor as is the third man along the conveyor-belt. Both will cause problems for those around them if they make errors. Many companies are now taking an active interest in the theory and practice of cyclic performance of employer and employee alike. In every case, improved performance has been recorded as a result, naturally leading to improved productivity.

Biorhythms and Air Travel

Nowhere has this been more obvious than in the accident statistics of airline companies in America. Records showed that 80 per cent of accidents occurred during domestic flights when either the pilot or co-pilot was in a critical stage of his rhythms. However, the biorhythms of air-traffic controllers were also analysed and the results so impressed many airline chiefs that few operatives now take sole command or fly if their biorhythms indicate the slightest chance of error. The reduction in accident figures has been significant as a consequence.

In cases where pilot error has appeared to be the only possible explanation for an incident, biorhythmic analyses have confirmed the prognoses. Where mechanical faults have seemed the most likely cause, researchers have been investigating the biorhythms of ground engineers for indications of error-prone phasing.

The overall accident rate is down by a not inconsiderable

percentage, although most airlines will not release their figures or admit to the use of biorhythms. Even though this is a long way away from the stage of universal acceptance of the use of biorhythms, it is a very reassuring fact for the average air passenger to know. And, while on the subject of travel, the introduction of a shift system which allowed public transport employees to be off the road or rails during the pertinent times must be a suggestion worthy of consideration.

Take the bus driver for example. Seated high in his cab, he is in a position to see far more than the public realize. Accidents involving buses do, of course, occur and sometimes the driver is at fault. But, if you take the time and trouble to observe them for a while, you will realize that bus-drivers spend much of their time avoiding accidents.

However, most of us are neither airline pilots nor bus-drivers and do not have to worry directly about the safety of the public. But we do all have to make simple decisions and emotional imbalance, physical inability or lack of stamina and vigour will affect us to varying degrees depending on our individual natures.

We cannot all emulate Superman or be as perfect as Batman on and off duty. We do not all have the patience of Job or the wisdom of Solomon. But all of us could, with a little thought, improve our way of life by having a quick look at our biocharts before embarking on the next project. There are some who would consider that this would be preordaining life and, in a limited sense, this may be so. But awareness of potential is one thing, utilizing it is quite another; as can be demonstrated by the following examples of notable achievements attained by remarkable men in various fields of endeavour.

Biorhythms in Action: Mark Spitz

Mark Spitz, who was born on 10 February 1950, achieved immortality when he created a record among record-breakers by winning seven gold medals at the 1972 Olympics at the end of August and the beginning of September.

Here we can see that, in fact, his biorhythms were at double peak form and must have contributed considerably towards his outstanding success. Between 27 August and 8 September,

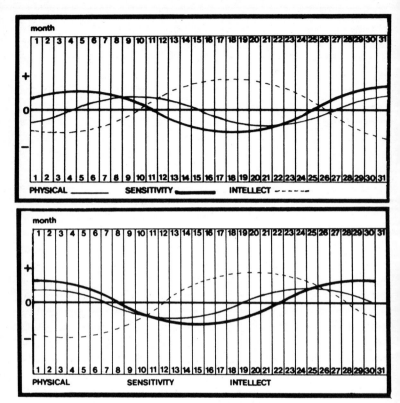

Figure 32: Mark Spitz

Spitz won medal after medal, almost within the space of a single week. On 27 August, he had a double critical in the physical and intellectual cycles and another double critical, this time in the physical and emotional cycles, on 8 September: surely a possible explanation for Spitz's enormous success.

Roger Bannister
Another remarkable athletic feat was achieved on 6 May 1954 at Oxford, England, when a young and popular athlete set out to break the four-minute mile with the help of willing

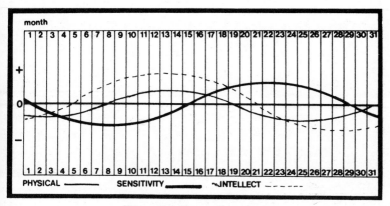

Figure 33: Roger Bannister

colleagues. Roger Bannister, born on 23 March 1929, had trained and planned hard for success, but his biorhythms were hardly auspicious for the occasion.

On the day, Bannister's physical and emotional rhythms were in negative phase, indicating a lack of stamina, and he had just completed an intellectual critical. He drove himself to exhaustion in the event and was almost on his knees when he crossed the tape in a remarkable 3 minutes 59.4 seconds. Had his physical cycle been up he might have clipped a few more points of a second off the record and not reduced himself to such a dangerously low state.

Neil Armstrong

Perhaps the most momentous occasion of the century did not take place on Earth, although it was a very special breed of earthman who stepped onto the surface of the Moon on 20 July 1969. Neil Armstrong, born 5 August 1930, set foot on the Moon with the immortal words: 'That's one small step for a man, one giant leap for mankind'.

Armstrong had blasted off on 16 July, a physically critical day, perhaps an inappropriate time for an event entailing such physical stress. He then experienced a triple mini-critical on the 22nd and had to undergo the enormous physical stress

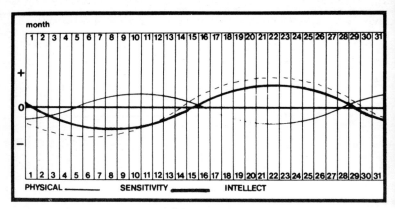

Figure 34. Neil Armstrong

of re-entry on the 24th. Despite the phasing of his physical cycle, Armstrong's two other rhythms were in peak form for balanced judgement. Remember, it was Armstrong who took over the controls and actually landed the lunar module. In those historic eight days, Armstrong's rhythms were just about perfect for such a task.

Because of the lack of gravity involved, he would not have required too much physical strength, as such, except for each end of the journey. However, a razor-sharp mind and emotional balance were essential for a successful mission. Here was something which no man had ever done before: just what would happen when he actually set foot on the Moon for the first time no one knew. Yet, in the event, it represents a case of perfect biorhythmic timing for what was, perhaps, the most daunting task ever undertaken by man.

On a more down to earth level, you can utilize your biorhythmic phasing in much the same way in everyday situations. You will experience an improvement in performance, achieve greater success and, if you time your efforts for best advantage, find life more rewarding, particularly if you also take the trouble to work out the cycles of others who are important to you.

At first, you will probably be over-conscious of your every move but, as time goes by and you get into the swing of things, the sensible utilization of biorhythms will become second nature. It is far easier to live in harmony with your biorhythms than it is to exist in discord with them.

8

CALCULATIONS

There are quite a few alternative methods of establishing the biorhythmic stages. These range from calculators that do the calculations for you in a matter of moments to having biorhythms charted for you by any one of several organizations. Besides these alternatives there are about three or four books that list predetermined codes which you simply look up in the tables provided. Unfortunately, however, these books are not readily available in this country.

The six or seven calculators available here are all more or less variations on a theme, and some of them will also automatically compute compatibilities. The price range starts at about £15 so, should you already own a pocket-calculator, it could prove an expensive exercise.

Applying for your biorhythm charts results in widely differing products. For a couple of pounds you may receive a computer print-out, in one of several various formats. The period covered can vary also; it may be anything from three months to a yearly readout: there appears to be no general conformity in this.

Visually, the easiest style to understand is the generally

agreed colour code: the physical cycle is always shown in red, the emotional cycle in blue and the intellectual in green. Each card is for one calendar month and all the biorhythms are easily and instantly recognizable. This means that you may travel anywhere in the world for any length of time and renew your biogram locally with confidence.

There are various alternative methods available ranging from a predetermined key code system that involves the physical and emotional cycles only, to a wristwatch which may be set with your personal rhythm and, once set, will remain constant.

All calculator and computer systems are limited to the current century, starting at 1 January 1901 through to 31 December 1999. This could make life a little difficult for students who wish to trace the lives of famous historical figures.

Establishing the Birth Date

However, with or without a calculator, it is possible to check the biorhythms for anyone whose birthdate is known in a few short minutes. This system may be confidently used for any date from 2 September 1752, when Great Britain decided to adopt the Gregorian Calendar in favour of the Julian. At that time, Britain was some eleven days adrift of other countries who had already made the change, therefore 2 September was followed immediately by 14 September. So, any date prior to this may be suspect, and you may need to verify your information before this particular period.

The American Colonies also changed on this date as they were subject to British law but other countries changed at different times before this and afterward. Generally speaking, most dates referred to in books have already been converted to the new system. A few may be quoted in the context of the calendar prevailing at the time in question but a visit to your local library will furnish you with the necessary details.

Students of history and historical figures will find it fascinating to delve into the pasts of their favourite heroes or villains and may well find that the biorhythms of these people could have had a bearing on the way they acted or thought at the time. A good biography often lists the full dates of an event so, armed with this and the birthdate of the individual concerned, an hour or two may be very pleasantly and profitably passed. The compatibility factor between well-known groups of people can also throw some interesting sidelights on their relationships.

A few minutes' work will give you similar results in respect of your modern favourites too, whether they be pop stars, actors or sportsmen. A newscaster may fluff his or her lines several times while you watch; a well known personality may act in a manner not normally associated with him; a golfer may seem to be completely off-form during a particular match; a whole host of occasions will offer you food for thought and provide the impetus necessary to check the biorhythms of your favourite personality.

First Steps

The basis of biorhythmic calculation is to determine the number of days that have elapsed since the date of birth up to the date you are interested in (remember to include the date of birth and the date in question in the total). This total is then divided by the number of days duration in each cycle: i.e. divide by 23 for the physical rhythm, 28 for the emotional and 33 for the intellectual cycle.

In each case the remainder figure will indicate the stage of the particular biorhythm under review; if there is no remainder this means that the day following is day one of a new positive phase, a critical day, the day on which this particular cycle is starting again.

In the physical rhythm, the positive phase starts on day one and lasts until day twelve which is a critical day. Days two and eleven are the plus stage which peaks at a mini-critical point on day seven. Days thirteen to twenty-three are the negative stage, with a mini-critical occurring on day eighteen.

In the emotional cycle, day one starts the positive phase, which lasts until day fifteen, the critical day, with a mini-critical on day eight; days two to fourteen are the plus phase. The negative stage runs from day sixteen to day twenty-eight, the mini-critical occurring on day twenty-two.

The intellectual rhythm starts on day one and is positive until the critical day, day seventeen; the mini-critical day occurs on day nine. The negative stage is from day eighteen until day thirty-three, with a mini-critical on day twenty-six.

Determining Biorhythms Without a Calculator

The tables in this section are devised to enable you to calculate biorhythms without a calculator. As an example we will take someone born on 24 May 1950 and calculate their biorhythms for 27 October 1979.

Figure 35 is to help you find your total number of days quickly. Where the specific number of years required is not listed, combine the necessary figures to obtain the desired total.

1 × 365 = 365	10 × 365 = 3650
2 × 365 = 730	20 × 365 = 7300
3 × 365 = 1095	30 × 365 = 10950
4 × 365 = 1460	40 × 365 = 14600
5 × 365 = 1825	50 × 365 = 18250
6 × 365 = 2190	60 × 365 = 21900
7 × 365 = 2555	70 × 365 = 25550
8 × 365 = 2920	80 × 365 = 29200
9 × 365 = 3285	90 × 365 = 32850

Figure 35

1756	1804	1852	1904	1952
1760	1808	1856	1908	1956
1764	1812	1860	1912	1960
1768	1816	1864	1916	1964
1772	1820	1868	1920	1968
1776	1824	1872	1924	1972
1780	1828	1876	1928	1976
1784	1832	1880	1932	1980
1788	1836	1884	1936	1984
1792	1840	1888	1940	1988
1796	1844	1892	1944	1992
	1848	1896	1948	1996

Figure 36: Leap Years Since 1752

Figure 37 (below) shows the number of each day throughout a
year, calculated progressively

Day	Jan.	Feb.	Mar.	Apr.	May	Jun.	Jul.	Aug.	Sep.	Oct.	Nov.	Dec.
1	1	32	60	91	121	152	182	213	244	274	305	335
2	2	33	61	92	122	153	183	214	245	275	306	336
3	3	34	62	93	123	154	184	215	246	276	307	337
4	4	35	63	94	124	155	185	216	247	277	308	338
5	5	36	64	95	125	156	186	217	248	278	309	339
6	6	37	65	96	126	157	187	218	249	279	310	340
7	7	38	66	97	127	158	188	219	250	280	311	341
8	8	39	67	98	128	159	189	220	251	281	312	342
9	9	40	68	99	129	160	190	221	252	282	313	343
10	10	41	69	100	130	161	191	222	253	283	314	344
11	11	42	70	101	131	162	192	223	254	284	315	345
12	12	43	71	102	132	163	193	224	255	285	316	346
13	13	44	72	103	133	164	194	225	256	286	317	347
14	14	45	73	104	134	165	195	226	257	287	318	348
15	15	46	74	105	135	166	196	227	258	288	319	349
16	16	47	75	106	136	167	197	228	259	289	320	350
17	17	48	76	107	137	168	198	229	260	290	321	351
18	18	49	77	108	138	169	199	230	261	291	322	352
19	19	50	78	109	139	170	200	231	262	292	323	353
20	20	51	79	110	140	171	201	232	263	293	324	354
21	21	52	80	111	141	172	202	233	264	294	325	355
22	22	53	81	112	142	173	203	234	265	295	326	356
23	23	54	82	113	143	174	204	235	266	296	327	357
24	24	55	83	114	144	175	205	236	267	297	328	358
25	25	56	84	115	145	176	206	237	268	298	329	359
26	26	57	85	116	146	177	207	238	269	299	330	360
27	27	58	86	117	147	178	208	239	270	300	331	361
28	28	59	87	118	148	179	209	240	271	301	332	362
29	29		88	119	149	180	210	241	272	302	333	363
30	30		89	120	150	181	211	242	273	303	334	364
31	31		90		151		212	243		304		365

On 27 October, our subject would have been twenty-nine
years old, plus the extra days. Therefore, from Figure 35:

$$20 \times 365 = 7300$$
$$+ \ \ 9 \times 365 = \underline{3285}$$
$$10585 \text{ (days). Carry forward.}$$

From Figure 36 we see that the subject has lived through
seven leap years, thus:

$$7$$
$$+ \ \underline{10585}$$
$$10592 \text{ (days). Carry this forward.}$$

We now need to know how many days have elapsed since the birthdate, 24 May 1950, until 27 October 1979. From Figure 37 we see that 27 October is day 300 and that 24 May is day 144, therefore:

$$
\begin{array}{r}
300 \\
-\ 144 \\
\hline
156
\end{array}
$$

Bring forward 10592 and add this to 156 to obtain the next figure required:

$$
\begin{array}{r}
10592 \\
+\ \ \ 156 \\
\hline
10748
\end{array}
$$

You must now add 1 to this figure to account for the day in question, thus:

$$
\begin{array}{r}
1 \\
+\ 10748 \\
\hline
\text{Total}\quad 10749
\end{array}
$$

Our subject has, therefore, lived a total of 10749 days, inclusive of the day for which we wish to calculate the biorhythms. To ascertain the physical rhythm we now divide this by 23. The whole figure will represent the number of full cycles experienced, the remainder figure will indicate the stage of the current cycle. Thus:

$$
23\ \overline{)\ 10749}\quad 467 \text{ remainder } 8
$$

Therefore, on 27 October the physical rhythm is 8 days into the plus phase.

Similarly, for the emotional rhythm we now divide 10749 by 28. Thus:

$$
28\ \overline{)\ 10749}\quad 383 \text{ remainder } 25
$$

On 27 October the emotional rhythm has reached day 25 of its cycle and is in negative phase.

Again, for the intellectual cycle we divide 10749 by 33. Thus:

$$
33\ \overline{)\ 10749}\quad 325 \text{ remainder } 24
$$

Therefore, on 27 October the intellectual rhythm is at day 24, in negative phase.

So, in the example used, the full biorhythmic reading for the day is: physical cycle day 8; emotional cycle day 25; intellectual cycle day 24.

The interpretation for this biorhythmic stage would be: an ideal day for clearing up anything which required physical participation or exertion but not too much concentration; mulling over a few minor problems while spring-cleaning or taking a nice long walk with the dog.

Determining Biorhythms Using a Calculator

In order to eliminate the mathematical drudgery, or to reduce the possibility of error, take an ordinary pocket-calculator and work out the number of days elapsed from the birthdate to the day in question utilizing Figures 35, 36 and 37. It is important to remember to add one day to your total to account for the day that you are working to. Using the same example as previously, your total will be 10749.

Using your calculator, divide this number by 23, 28 and 33. Thus:

$$10749 \div 23 = 467.34782 \text{ (physical cycle)}$$
$$10749 \div 28 = 383.89285 \text{ (emotional cycle)}$$
$$10749 \div 33 = 325.72727 \text{ (intellectual cycle)}$$

For the physical biorhythm stage, multiply the decimal remainder by 23, thus: $.34782 \times 23 = 7.99986$. To the nearest whole number this is 8.

For the emotional biorhythm stage, multiply the decimal remainder by 28, thus: $.89285 \times 28 = 24.9998$. To the nearest whole number this is 25.

For the intellectual biorhythm stage, multiply the remainder by 33, thus: $.72727 \times 33 = 23.99991$. To the nearest whole number this is 24.

Therefore, the biorhythms for the day are: physical cycle, day 8; emotional cycle, day 25; intellectual cycle, day 24.

Once you have calculated the figures of these cycles for any particular date it is a simple matter to make up your own biogram for a month. Whether you wish to go back in time to see why performance was impaired or better than average, or whether you wish to make up a card for a month in advance, this will provide you with two very simple methods of producing biorhythm charts without the aid of a biorhythmic calculator.

All the information required for calculating and utilizing biorhythms has been provided in this book, and there are a number of blank charts at the back for you to experiment with (see p.127). All you need, therefore, is three coloured pens or pencils and a small protractor or other regular curved object to complete the graphs.

If you do your calculations for the first day of any month in order to arrive at the starting figures in each of the three cycles, it will be easier to pick out the critical days, draw the curves between these points and complete the rest of the month accordingly. But do check carefully to ensure that you have got everything right, particularly at first: with practice you will quickly get the hang of it.

Here is another example, just to help you get into the swing of things. The working has been approached from a slightly different angle this time.

We will assume that we want the biorhythms for 24 June 1979 for someone born 1 October 1938. On 1 October 1979 our subject would be forty-one years old; therefore, the whole calculation is: 40 × 356 + leap years + days from 1 October 1978 + 1.

From Figure 35 we get: 40 × 365 = 14600

From Figure 36, leap years from 1938 until 1979 = 10, thus:

$$
\begin{array}{r}
14600 \\
+ \quad 10 \\
\hline
14610
\end{array}
$$

From Figure 37 we find that 1 October is day 274 and 24 June is day 175. Counting forward, there are 91 days to the end of the year, plus 175 days to 24 June of the following year. Therefore:

$$
\begin{array}{r}
91 \\
+ \quad 175 \\
\hline
266
\end{array}
$$

Thus:

$$
\begin{array}{r}
266 \\
+ \ 14610 \\
\hline
14876 \\
+ \quad 1 \ \text{(for the day in question)} \\
\hline
14877
\end{array}
$$

We now divide this total figure by the relative cycle lengths: 23 for the physical, 28 for the emotional and 33 for the intellectual. We will do this first without a calculator.

Physical:

$$23 \overline{\smash{)}\,14877} \quad 646 \text{ remainder } 19$$

Emotional:

$$28 \overline{\smash{)}\,14877} \quad 531 \text{ remainder } 9$$

Intellectual:

$$33 \overline{\smash{)}\,14877} \quad 450 \text{ remainder } 27$$

The biorhythms for the day in question are, therefore: physical cycle, day 19; emotional cycle, day 9; and intellectual cycle, day 27.

If we use a pocket-calculator we get a figure of 14877 as the number of days elapsed. We now divide this total by the appropriate cycle lengths:

Physical: $14877 \div 23 = 646.82608$
Emotional: $14877 \div 28 = 531.32142$
Intellectual: $14877 \div 33 = 450.81818$

Now multiply the decimal remainder by the cycle length involved:

Physical: $.82608 \times 23 = 18.99984$ (to the nearest whole number, this gives 19)
Emotional: $.32142 \times 28 = 8.99976$ (to the nearest whole number, this gives 9)

Intellectual: .81818 × 33 = 26.99994 (to the nearest whole number, this gives 27)

The biorhythms for the day are, therefore:

Physical cycle, day 19; emotional cycle, day 9; and intellectual cycle, day 27.

Biorhythmically, this would be interpreted as a good day for getting on with people in general, but a day on which to avoid committing oneself to sudden or important decisions without serious forethought.

After a few practice runs, you should find it fairly easy to work out these calculations for yourself, your friends and relatives, or anyone who attracts your interest at the moment.

In the Appendix of this book are the birthdates of over 300 well-known people from all walks of life: stage, history, sport, politics past and present, and those individuals who have made a niche for themselves in the pages of history by their actions, words or thoughts.

As long as you remember that biorhythms do not in themselves have a cause and effect but are subject to prevailing conditions, you will find that their stage and phase may have had a bearing on why an individual behaved in a certain manner on a particular occasion. For example, it is often a source of surprise to discover the basis of famous partnerships: biorhythmically speaking.

But, where *you* are concerned – and that is what this book has been all about – learn to live with your biorhythms in order to experience a more satisfactory way of life; in every sense of the word.

APPENDIX

CELEBRITY BIRTHDATES

Note: Every effort has been made to ensure that the following list is accurate and inclusion here means that at least three different published sources have been consulted for verification.

Dawn Addams	21 September 1930
Louisa M. Alcott	29 November 1832
Princess Alexandra	25 December 1936
Muhammed Ali	18 January 1942
Woody Allen	1 December 1935
Ursula Andress	19 March 1938
Prince Andrew	19 February 1960
Eamonn Andrews	19 December 1922
Julie Andrews	1 October 1935
Paul Anka	30 July 1941
Princess Anne	15 August 1950
Anthony Armstrong-Jones	7 March 1930
Arthur Askey	6 June 1900
Fred Astaire	10 May 1899
Clement Attlee	3 January 1883

Lauren Bacall	16 September 1924
Burt Bacharach	12 May 1929
Lord Baden-Powell	22 February 1857
Douglas Bader	21 February 1910
Joan Baez	19 April 1941
Anne Bancroft	17 September 1931
Gene Barry	4 June 1922
Sir Thomas Beecham	29 April 1879
Alexander Graham Bell	3 March 1847
Tony Bennett	3 August 1926
Ingrid Bergman	29 August 1917
Bill Bixby	22 January 1934
Victor Borge	3 January 1909
Patti Boulaye	3 March 1954
David Bowie	8 January 1947
Charles Bronson	3 November 1922
Richard Burton	10 November 1925
James Caan	26 March 1939
Michael Caine	14 March 1933
James Callaghan	27 March 1912
Jimmy Carter	1 October 1924
Pablo Casals	29 December 1876
Barbara Castle	6 October 1911
Fidel Castro	13 August 1926
Neville Chamberlain	18 March 1869
Carol Channing	31 January 1923
Charles Chaplin	16 April 1889
Prince Charles	14 November 1948
Lorraine Chase	16 July 1951
Sean Connery	25 August 1930
Jimmy Connors	2 September 1952
Henry Cooper	3 May 1934
Ronnie Corbett	4 December 1930
Marie Curie	7 November 1867
Salvador Dali	11 May 1905
Bette Davis	5 April 1908
Sammy Davis Junr	8 December 1925
Doris Day	3 April 1924

Sandra Dee	23 April 1942
John Denver	31 December 1943
Charles Dickens	7 February 1812
Angie Dickinson	30 September 1931
Walt Disney	5 December 1901
Kirk Douglas	9 December 1916
Sir Arthur Conan Doyle	22 May 1859
Faye Dunnaway	14 January 1941
Bob Dylan	24 May 1941
Clint Eastwood	31 May 1931
Mary Baker Eddy	16 July 1821
Sir Anthony Eden	12 June 1897
Vincent Edwards	7 July 1928
Samantha Eggar	3 May 1939
Albert Einstein	14 March 1879
Dwight Eisenhower	14 October 1890
Anita Ekberg	29 September 1931
Queen Elizabeth II	21 April 1926
David Essex	23 July 1948
Kenny Everett	25 December 1944
Chris Evert	21 December 1954
Peter Falk	16 September 1927
Jose Feliciano	10 September 1945
Roberta Flack	10 February 1940
Henry Fonda	16 May 1905
Jane Fonda	21 December 1937
Joan Fontaine	22 October 1917
Dame Margo Fonteyn	18 May 1919
Gerald Ford	14 July 1913
Glen Ford	1 May 1916
General Franco	4 December 1892
Aretha Franklin	25 March 1942
Lady Antonia Fraser	27 August 1932
Clement Freud	24 April 1924
Sigmund Freud	6 May 1856
David Frost	7 April 1939
Sir Vivian Fuchs	11 February 1913

Yuri Gagarin	9 March 1934
Greta Garbo	18 September 1905
Art Garfunkel	19 October 1942
Judy Garland	10 June 1922
James Garner	7 April 1928
General de Gaulle	22 November 1890
Paul Getty	15 December 1892
John Glenn	18 July 1921
Herman Goering	12 January 1893
Billy Graham	7 November 1918
Cary Grant	18 January 1904
Larry Grayson	31 August 1923
Germaine Greer	29 January 1939
Jo Grimmond	29 July 1921
Sir Alec Guiness	2 April 1914
Gene Hackman	30 January 1931
Antony Hancock	3 May 1924
Valerie Harper	22 August 1940
George Harrison	25 February 1943
Rex Harrison	5 March 1908
Goldie Hawn	21 November 1945
Edward Heath	9 July 1916
Audrey Hepburn	4 May 1929
Sir Edmund Hillary	20 July 1919
Alfred Hitchcock	13 August 1899
Adolf Hitler	20 April 1889
Lew Hoad	23 November 1934
Dustin Hoffman	8 August 1937
Bob Hope	29 May 1903
Harry Houdini	6 April 1874
Sir Geoffrey Howe	20 December 1926
Engelbert Humperdink	2 May 1936
Sir Len Hutton	23 June 1916
Jaques Ibert	15 August 1890
Henrik Ibsen	20 March 1828
Vincent d'Indy	27 March 1851
Samual Ingersoll	3 March 1818
Shah of Iran	26 October 1919

Mick Jagger	26 July 1943
Glenda Jackson	9 May 1936
David Janssen	27 March 1930
Ingemar Johansson	22 September 1932
Elton John	25 March 1947
Al Jolson	26 May 1886
Tom Jones	7 June 1940
Sir Keith Joseph	17 January 1918
Carl Jung	26 July 1875
Danny Kaye	18 January 1913
Stacey Keach	2 June 1941
Helen Keller	27 June 1880
Gene Kelly	23 August 1912
Grace Kelly	12 November 1929
Edward Kennedy	22 February 1932
Deborah Kerr	30 September 1921
Billie Jean King	22 November 1943
Henry Kissinger	27 May 1923
Lord Kitchener	24 June 1850
Evel Knievel	17 October 1938
Gladys Knight	28 May 1944
Kris Kristofferson	22 June 1936
Jim Laker	9 February 1922
Dorothy Lamour	10 October 1914
Burt Lancaster	2 November 1913
Lillie Langtry	13 October 1854
Angela Lansbury	16 October 1925
T.E. Lawrence	15 August 1888
Christopher Lee	27 May 1922
John Lennon	9 October 1940
Liberace	16 May 1919
Abraham Lincoln	12 February 1809
Charles Lindbergh	4 February 1902
Franz Liszt	22 October 1811
Jack Lord	30 December 1934
Sophia Loren	20 September 1934
Joe Louis	13 May 1914

Ali MacGraw	1 April 1939
Shirley MacLaine	24 April 1934
Harold Macmillan	10 February 1894
Paul McCartney	18 June 1942
Karl Malden	22 March 1914
Princess Margaret	21 August 1930
Golda Meir	3 May 1898
Bob Monkhouse	1 June 1928
Anne Moore	20 August 1950
Patrick Moore	4 March 1923
Roger Moore	14 October 1927
Eric Morecambe	14 May 1926
Oswald Mosley	16 November 1896
Field Marshall Montgomery	17 November 1887
Earl Mountbatten	25 June 1900
Audie Murphy	20 June 1924
Pete Murray	19 September 1925
Ilie Nastase	19 July 1946
Dame Anna Neagle	20 October 1904
Patricia Neal	20 January 1926
Anthony Newley	24 September 1931
Paul Newman	26 January 1925
Jack Nicklaus	21 January 1940
Vaslaw Nijinsky	28 February 1890
Leonard Nimoy	26 March 1931
Richard Nixon	9 January 1913
Dennis Norden	6 February 1922
Barry Norman	21 August 1933
Kim Novak	13 February 1933
Rudolf Nureyev	17 March 1938
Merle Oberon	19 February 1911
Hugh O'Brian	19 April 1930
Maureen O'Hara	17 August 1921
Sir Laurence Olivier	22 May 1907
Jackie Onassis	28 July 1929
Ryan O'Neal	20 April 1949
George Orwell	25 June 1903
John Osborne	12 December 1929

Tessie O'Shea	13 March 1914
Donny Osmond	9 December 1957
Peter O'Toole	2 August 1933
Al Pacino	25 April 1940
Arnold Palmer	10 September 1929
Dolly Parton	19 January 1946
Robert Peary	6 May 1856
George Peppard	1 October 1933
Prince Philip	10 June 1921
Pablo Picasso	25 October 1881
Harold Pinter	10 October 1930
Edgar Allen Poe	19 January 1809
Sidney Poitier	20 February 1927
Roman Polanski	18 August 1933
Enoch Powell	16 June 1912
Otto Preminger	5 December 1906
Andre Previn	6 April 1929
Vincent Price	27 May 1911
James Prior	11 October 1927
Mary Quant	11 February 1934
Anthony Quinn	21 April 1916
Anthony Quayle	7 September 1913
Roger Quilter	1 November 1877
Terence Rattigan	10 June 1911
Ronald Reagan	6 February 1911
Helen Reddy	25 October 1941
Robert Redford	18 August 1937
Vanessa Redgrave	30 January 1937
Lee Remick	14 December 1935
Diana Rigg	20 July 1938
Robespierre	6 May 1758
Ginger Rogers	16 July 1911
Erwin Rommel	15 November 1891
Mickey Rooney	23 September 1920
Theodore Roosevelt	27 October 1858
Diana Ross	26 March 1944

Sir Malcolm Sargent	29 April 1895
Telly Savalas	21 January 1924
Peter Sellers	8 September 1925
Ernest Shackleton	15 February 1874
Dinah Shore	1 March 1921
Carly Simon	25 June 1945
Frank Sinatra	12 December 1915
Cyril Smith	28 June 1928
Ian Smith	8 April 1919
Pat Smythe	22 November 1928
Boris Spassky	30 January 1937
Dr Benjamin Spock	3 May 1903
Barbara Stanwyck	16 July 1907
Ringo Starr	7 July 1940
Tommy Steele	17 December 1936
Barbra Streisand	24 April 1942
Gloria Swanson	27 March 1899
Margaret Thatcher	13 October 1925
Terry Thomas	14 July 1911
Jeremy Thorpe	29 April 1929
Gene Tierney	20 November 1920
Marshall Tito	25 May 1892
Mel Torme	13 September 1925
Paul Tortelier	21 March 1914
Tommy Trinder	24 March 1909
Sophie Tucker	13 January 1884
Gene Tunney	25 May 1898
Lana Turner	8 February 1920
Rita Tushingham	14 March 1942
Twiggy	19 September 1949
Leslie Uggams	25 May 1943
Ulanova	10 January 1910
Liv Ullman	16 December 1939
Stanley Unwin	7 June 1911
Mary Ure	18 February 1933
Peter Ustinov	16 April 1921
Maurice Utrillo	26 December 1883

Dick Van Dyke	13 December 1925
Rudolph Valentino	6 May 1895
Frankie Vaughan	3 February 1928
Robert Vaughan	22 November 1932
Sarah Vaughan	27 March 1924
Jules Verne	8 February 1928
Gore Vidal	3 October 1925
Lindsay Wagner	22 June 1949
Clint Walker	30 May 1927
George C. Wallace	25 August 1919
Sir Barnes Wallis	26 September 1887
John Wayne	26 May 1907
Raquel Welch	5 September 1942
Tom Weiskopf	9 November 1942
Orson Welles	6 May 1915
Mae West	17 August 1893
William Whitelaw	28 June 1918
Oscar Wilde	16 October 1854
Kenneth Williams	22 February 1926
Sir Harold Wilson	11 March 1916
Duke of Windsor	23 June 1894
Henry Winkler	30 October 1945
Ernie Wise	27 November 1925
Stevie Wonder	13 May 1950
Michael York	27 March 1942
Susannah York	9 January 1942
Andrew Young	12 March 1922
Gig Young	4 November 1917
Jimmy Young	21 September 1921
Loretta Young	6 January 1913
Darryl Zanuck	5 September 1902
Franco Zefferelli	12 February 1923
Mai Zetterling	24 May 1925
Ferdinand Zeppelin	8 July 1838
Florenz Ziegfeld	21 March 1869
Anne Ziegler	22 June 1910
Efrem Zimbalist jnr	30 November 1923
Emile Zola	2 April 1840
Pinchas Zuckerman	16 July 1948

INDEX

BLANK BIORHYTHM CHARTS

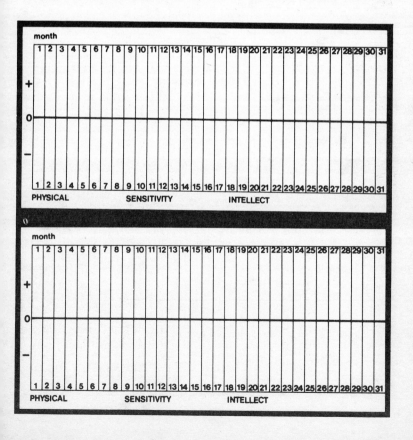

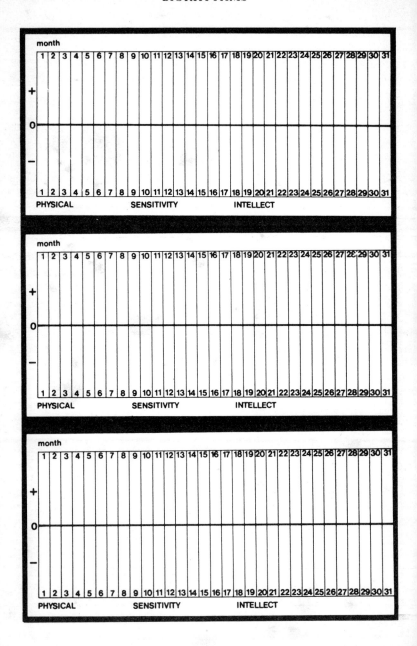